D0121533

Integrating Information Infrastructure with Geographical Information Technology

To HMDG

INNOVATIONS IN GIS 6

Integrating Information Infrastructures with Geographical Information Technology

Edited by
Bruce M. Gittings
Department of Geography,
The University of Edinburgh.

UK   Taylor & Francis Ltd, 11 New Fetter Lane, London EC4P 4EE
USA   Taylor & Francis Inc., 325 Chestnut Street, Philadelphia PA 19106

Taylor & Francis is a member of the Taylor & Francis group

Printed and bound in Great Britain by TJ International Ltd, Padstow, Cornwall

British Library Cataloguing in Publication Data

A catalogue record for this book is available from the British Library.
ISBN 0-7484-0886-X (cased)

Library of Congress Cataloguing in Publication Data are available

Contents

Preface

This volume represents the latest in a series, first published in 1994, based on the GIS Research UK (GISRUK) conference series initiated by Professor Mike Worboys in 1993. As originally conceived, the *Innovations* series was intended to provide a window onto the research presented at these GISRUK conferences, by including a selection of the papers, which were very much intended to cover the breadth of GIS research. In the intervening five years, much has been achieved. Geographical Information Systems have quite rightly evolved into a Geographical Information Science; both the technology and the theory which form this science have become more robust, yet alongside this, the technology has also become more accessible, with client-based tools becoming well established. Much of GIS has moved from project-based to corporate implementation, with significant implications for data management. Lastly, the World Wide Web has appeared, as if from nowhere.

Thus, it is appropriate to take time for reflection and consider how the *Innovations* series should develop over the next five years. We need to decide whether it remains appropriate to contemplate a volume which acts as an extension to a conference proceedings, covering the entirety of GIS. The answer, of course, is no; the success of what we have done in past years has been such that the field has both broadened and deepened. While the innovative research being reported has grown multi-fold, the diversification of the field means that, quite appropriately, most will be reported in an increasing number of themed journals or other specialist publications. These publications are focussed towards the particular interests, whether theoretical or application-oriented, of the ever-increasing number of GIS researchers. Thus, the time has come to move away from the 'book of the conference'. It is our intention that each volume should now have a life of its own, rather than simply act as a historical record of the host of different topics which have been presented at the GISRUK conferences.

To achieve this aim, I have chosen the theme represented by the title of this volume, which I feel is not only one of current importance, but I would regard as being critical to the future success and uptake of GIS. I have selected chapters based on their relevance to the theme, and clearly this is a significant departure from what has gone before. Authors have been asked not to simply reproduce their conference papers, but to augment and extend their ideas in the context of the theme. The chapters have all been thoroughly edited. Authors have been able to take into account the comments received at the conference and also those subsequently given by referees. Through this rigorous, but supportive, assistance from referees I have been able to shape a book which develops its theme through a logical progression of issues and ideas, which are very much related. Thus, what you have in your hands is the first of a new-look *Innovations* series, which I hope is primed for a new millennium.

However, it would be foolish to abandon the many successful features of the series to date. And indeed I must take account of the excellent lead given by previous editors. The research reported in past *Innovations* has always been outward looking, beyond the shores of the United Kingdom, taking into account the many international researchers who have attended GISRUK conferences. This has not only given the *Innovations* series an international appeal, but ensures we do not become parochial in our outlook, giving cognicance to achievements, and lessons learned, elsewhere.

This volume has been possible because of the continued success of the GISRUK conference series. Because of the diversity of excellent quality research presented, we are able to make the selection for *Innovations* on a thematic basis. Thus the success of GISRUK is tied closely to the success of the *Innovations* series

Turning to the most recent conference, GISRUK '98, in Edinburgh, attracted almost 200 delegates, importantly these delegates came from not just the UK, but many European countries and elsewhere, ensuring GISRUK is maintained as a supra-national conference, which adds greatly to its value. The GISRUK community cuts across traditional discipline boundaries; although still attracting a majority of geographers, significant minorities come from Archaeology, Environmental Science, Geomatics, Computer Science and Geology.

At this point, it is worth reviewing the aims which we have set out for the GISRUK conference series:

* to act as a focus for GIS research in the UK
* to provide a mechanism for the announcement and publication of GIS research
* to act as an interdisciplinary forum for the discussion of research ideas
* to promote active collaboration amongst researchers from diverse parent disciplines
* to provide a framework in which postgraduate students can see their work in a national context

I am convinced that we continue to be successful on all counts!

A feature of GISRUK'98 has been increasing attendance from those beyond the academic research community; we are embracing individuals in commerce and industry, who value the free-flowing ideas, and debates, that have become the trademark of the GISRUK conference series. Provided GISRUK maintains its research-based focus, this is a healthy development.

The conference has developed a 'personality' based on a unique blend of panel-led debates, round-table discussion sessions, poster papers, software demonstrations, a Publishers Exhibition, all combined with an intensive social programme which gives a friendly and open forum for the exchange of ideas. GISRUK'98 included a formal debate on the UK National Geospatial Data Framework (NGDF), with an international contribution provided by Nancy Tosta.

The round-table sessions allow free-flowing discussion on a range of research foci. For 1998, these included a critical review of object-oriented GIS, digital terrain modelling, GIS in support of traffic and transport planning, future census geography, web-based GIS, the changing needs of the GIS industry (education, training, employment and professionalism), GIS and health, scale issues, temporal GIS, decision support systems in environmental planning and management.

A new innovation was a prize for the best presentation by a young researcher. The prize was the suggestion of, and generously donated by, Robin McLaren of Know Edge Ltd. Nathan Williams won the prize, and his paper has been developed into his chapter in this volume. The focus on youth and new ideas is developed through the close association between GISRUK and the Young Researcher's Forum, organised by RRLnet. However, we as a steering committee welcome your ideas and involvement. GISRUK benefits from having no organisation running it, and therefore is able to directly reflect the needs and views of the community it serves. Without you, the researchers, presenters and contributors we would have nothing to discuss, and indeed no conference and no need for books!

Acknowledgements

There are many people who have contributed to this book and to the GISRUK'98 conference, and I must take this opportunity to thank them all. Taking on the organisation of a conference is not something I would recommend; to then continue to edit a book is verging on madness! However, the support I have received from all involved has been remarkable, and has made both duties much simpler than they would otherwise have been.

All of the authors who have contributed to this volume, whose names are laid out in the following section, have worked extremely hard to prepare and refine chapters. In many cases this has involved extension or reworking of the ideas they presented at the original conference to fully embrace the theme of this book. Particular mention must be made of Nancy Tosta and Menno-Jan Kraak who have had the difficult job of putting their work into a national context which is not their own; both have successfully given us a range of stimulating ideas.

I am grateful also to William Mackaness and Neil Stuart, who offered help and encouragement at exactly the right time, and to Hermione Cockburn who ably assisted with the editorial work. Both Tony Moore and Luke Hacker, at Taylor and Francis, provided much help at various stages in the realisation of this volume. Colleagues on the GISRUK steering committee offered advice and encouragement throughout this project and the organisation of the conference. I hope I can reciprocate, to the same extent, in the future.

Turning to the GISRUK'98 conference, we are all grateful to the various companies and organisations who sponsored the conference. Their confidence in supporting this event shows a clear commitment to nurturing research and to the broader GI profession.

I am enormously grateful to Andy Lewis, Emma Sutton and Duncan Moss, who assisted beyond the call of duty, together with an eager band of student helpers. Sandra Cowie and her staff at the Lister Institute, and particularly my colleagues on the local organising committee, for a wide range of help from last-minute reviewing to ensuring various aspects of the conference ran smoothly.

On behalf of the Steering Committee I would like to thank David Parker, who has steadfastly chaired our Steering Committee through the last few years, and following much hard work and a well-deserved promotion has stepped aside. We should also record our thanks Mike Worboys, who perceived the need for these conferences, ran the first of the series in 1993 and set an excellent model with Innovations in GIS 1.

Bruce M. Gittings,
Edinburgh, 1998.

Contributors

Richard H. A. Baker
MAFF Central Science Laboratory, Sand Hutton, York YO41 1LZ, United Kingdom.
E-mail: r.baker@csl.gov.uk

Andre Bargiela
Department of Computing, The Nottingham Trent University, Newton Building, Burton Street, Nottingham NG1 4BU, United Kingdom.

Jim Barton
Department of Physics, Heriot-Watt University, Riccarton, Edinburgh EH14 4AS, United Kingdom.
E-mail: jsb@phy.hw.ac.uk

Marianne L. Broadgate
Department of Geology and Geophysics, University of Edinburgh, The King's Buildings, West Mains Road, Edinburgh EH9 3JW, United Kingdom.
E-mail: Marianne.Broadgate@ed.ac.uk

Christophe Claramunt
Department of Computing, The Nottingham Trent University, Newton Building, Burton Street, Nottingham NG1 4BU, United Kingdom.
E-mail: clac@doc.ntu.ac.uk

John Davy
School of Geography, University of Leeds, Leeds LS2 9JT, United Kingdom.

Martin Dodge
Centre for Advanced Spatial Analysis, University College London, Gower Street, London WC1E 6BT, United Kingdom.

Mark Dorey
School of Computing, University of Glamorgan, Treforrest, Pontypridd, Mid Glamorgan CF37 1Dl, United Kingdom.
E-mail: midorey@glam.ac.uk

Jane Drummond
Department of Geography and Topographic Science, University of Glasgow, Glasgow G12 8QQ, United Kingdom.
E-mail: jdrummond@geog.gla.ac.uk

Matt Duckham
Department of Geography and Topographic Science, University of Glasgow, Glasgow G12 8QQ, United Kingdom.
E-mail: mduckham@geog.gla.ac.uk

Roger Dunham
Macaulay Land Use Research Institute, Craigiebuckler, Aberdeen AB15 8QH, United Kingdom.
E-mail: r.dunham@mluri.sari.ac.uk

Adam Etches
Department of Computing, The Nottingham Trent University, Newton Building, Burton Street, Nottingham NG1 4BU, United Kingdom.

Zhiqiang Feng
Department of Geography, Lancaster University, Lancaster LA1 4YB, United Kingdom.
E-mail: z.feng@lancaster.ac.uk

Robin Flowerdew
Department of Geography, Lancaster University, Lancaster LA1 4YB, United Kingdom.
E-mail: r.flowerdew@lancaster.ac.uk

Bruce M. Gittings
Department of Geography, University of Edinburgh, Drummond Street, Edinburgh EH8 9XP, United Kingdom.
E-mail: bruce@geo.ed.ac.uk

Jasmee Jaafar
School of Geography, University of Nottingham, University Park, Nottingham NG7 2RD, United Kingdom.
E-mail: jaafar@geography.nottingham.ac.uk

Claire H. Jarvis
MAFF Central Science Laboratory, Sand Hutton, York YO41 1LZ, United Kingdom.
E-mail: chj@geo.ac.ed.uk

Christopher Jones
School of Computing, University of Glamorgan, Treforrest, Pontypridd, Mid Glamorgan CF37 1Dl, United Kingdom.
E-mail: cbjones@glam.ac.uk

Zarine Kemp
Computing Laboratory, University of Kent at Canterbury, Canterbury CT2 7NF, United Kingdom.
E-mail: zk@ukc.ac.uk

David Kidner
School of Computing, University of Glamorgan, Treforrest, Pontypridd, Mid Glamorgan CF37 1Dl, United Kingdom.
E-mail: dbkidner@glamorgan.ac.uk

Iisakki Kosone
Department of Computing, The Nottingham Trent University, Newton Building, Burton Street, Nottingham NG1 4BU, United Kingdom.

Menno-Jan Kraak
ITC, Department of Geoinformatics, P.O. Box 6 - 7500 AA Enschede, The Netherlands.
E-mail: kraak@itc.nl

Stephen McAllister
Faculty of informatics, University of Ulster at Coleraine, Cromore Road, Coleraine, Co. Londonderry BT52 1SA, United Kingdom.
E-mail: JMS.McAlister@ulst.ac.uk

Donna McCormick
School of Environmental Studies, University of Ulster at Coleraine, Cromore Road, Coleraine, Co. Londonderry BT52 1SA, United Kingdom.
E-mail: D.McCormick@ulst.ac.uk

James Macgill
Centre for Computational Geography, School of Geography, University of Leeds, Leeds LS2 9JT, United Kingdom.
E-mail: j.macgill@geography.leeds.ac.uk

William Mackaness
Department of Geography, University of Edinburgh, Drummond Street, Edinburgh EH8 9XP, United Kingdom.
E-mail: wam@geo.ed.ac.uk

David R. Miller
Macaulay Land Use Research Institute, Craigiebuckler, Aberdeen AB15 8QH, United Kingdom.
E-mail: d.miller@mluri.sari.ac.uk

Adrian Moore
School of Environmental Sciences, University of Ulster at Coleraine, Cromore Road, Coleraine, Co. Londonderry BT52 1SA, United Kingdom.
E-mail: a.moore@ulst.ac.uk

Derek Morgan
MAFF Central Science Laboratory, Sand Hutton, York YO41 1LZ, United Kingdom.
E-mail: d.morgan@cls.gov.uk

Stan Openshaw
School of Geography, University of Leeds, Leeds LS2 9JT, United Kingdom.
E-mail: stan@geog.leeeds.ac.uk

Gerard Parr
*Faculty of Informatics, University of Ulster at Coleraine, Cromore Road, Coleraine,
Co. Londonderry BT52 1SA, United Kingdom.*
E-mail: gp.parr@ulst.ac.uk

Carsten Peter
IVU Umwelt, Burgweg 10, 79350 Sexau, Germany.
E-mail: cap@ivu.de

Anup Pradhan
*Department of Geography, University of Edinburgh, Drummond Street, Edinburgh
EH8 9XP, United Kingdom.*
E-mail: anp@geo.ed.ac.uk

Gary Priestnall
*School of Geography, University of Nottingham, University Park, Nottingham NG7 2RD,
United Kingdom.*
E-mail: Gary.Priestnall@nottingham.ac.uk

Ross Purves
*Department of Geography, University of Edinburgh, Drummond Street, Edinburgh
EH8 9XP, United Kingdom.*
E-mail: rsp@geo.ed.ac.uk

Phillip J. Rallings
*Division of Mathematics and Computing, University of Glamorgan, Treforrest,
Pontypridd, Mid Glamorgan CF37 1DL, United Kingdom.*
E-mail: pjrallin@glam.ac.uk

Narushige Shiode
*Centre for Advanced Spatial Analysis, University College London, Gower Street, London
WC1E 6BT, United Kingdom.*
E-mail: nshiode@geography.ucl.ac.uk

Cláudio de Souza Baptista
*Computing Laboratory, University of Kent at Canterbury, Canterbury CT2 7NF,
United Kingdom.*
E-mail: cdsb1@ukc.ac.uk

Andrew Sparkes
*School of Computing, University of Glamorgan, Treforrest, Pontypridd, Mid Glamorgan
CF37 1Dl, United Kingdom.*

Neil Stuart
Department of Geography, University of Edinburgh, Drummond Street, Edinburgh
EH8 9XP, United Kingdom.
E-mail: ns@geo.ed.ac.uk

David E. Sugden
Department of Geography, University of Edinburgh, Drummond Street, Edinburgh
EH8 9XP, United Kingdom.
E-mail: des@geo.ed.ac.uk

Nancy Tosta
Director of Forecasting and Growth Strategy, Puget Sound Regional Council, Seattle,
Washington, USA.
E-mail: NTosta@psrc.org

Ian Turton
Centre for Computational Geography, School of Geography, University of Leeds, Leeds
LS2 9JT, United Kingdom.
E-mail: ian@geography.leeds.ac.uk

Andrew Ware
Division of Mathematics and Computing, University of Glamorgan, Treforrest,
Pontypridd, Mid Glamorgan CF37 1Dl, United Kingdom.
E-mail: jaware@glam.ac.uk

Mark Ware
School of Computing, University of Glamorgan, Treforrest, Pontypridd, Mid Glamorgan
CF37 1Dl, United Kingdom.
E-mail: jmware@glam.ac.uk

Nathan A. Williams
Department of Geography, Virtual Field Course/Virtual Reality GIS Research, Birkbeck
College, University of London, 7-15 Gresse St., London W1P 2LL, United Kingdom.
E-mail: n.williams@geog.bbk.ac.uk

GISRUK Committees

GISRUK National Steering Committee

Steve Carver	University of Leeds
Jane Drummond	University of Glasgow
Bruce Gittings (Chair)	University of Edinburgh
Peter Halls	University of York
Gary Higgs	University of Wales at Cardiff
Zarine Kemp	University of Kent
David Kidner	University of Glamorgan
David Martin	University of Southampton
George Taylor	University of Newcastle

GISRUK '98 Local Organising Committee

Bruce Gittings (Chair)	Sandra Cowie	Steve Dowers
Bob Hodgart	Nick Hulton	Roger Kirby
Andy Lewis	William Mackaness	David Macleod
Tim Malthus	Chris Place	Neil Stuart
Fran Taylor	Alistair Towers	

GISRUK '98 Sponsors

The GISRUK Steering Committee are enormously grateful to the following organisations who generously supported GISRUK '98:

The Association for Geographic Information (AGI)
Taylor & Francis Ltd
CACI Ltd
ESRI (UK) Ltd
Experian Ltd
GeoInformation International
Know Edge Ltd
National Geospatial Data Framework
Oxford University Press
PCI Geomatics Ltd
SmallWorld
Whitbread plc

1

Introduction

Bruce M. Gittings

1.1 INTEGRATING INFORMATION INFRASTRUCTURES WITH GEOGRAPHICAL INFORMATION TECHNOLOGY

It is rare to have the opportunity for reflection; to learn from the past, but perhaps more interestingly, to speculate as to the future. The pleasure of editing this volume is that I can do just that; through the considered selection of chapters, I have been able to draw on a set of themes, and associated issues, which provide an essentially personal view of the state of GI research and technology. Through this volume, two common themes will be echoed; namely, turning data into useful information and the importance of new technology.

The most significant technological theme, which will be explored, is that of the World Wide Web (WWW). Built on the somewhat inadequate foundations of the Internet, the Web has come from nothing just five years ago, to represent the vast majority of Internet traffic today. Indeed, the Web has *become* the Internet in the minds of many. Yet, there are many challenges to be faced. Buttenfield (1997) reviews some of the political and cultural issues relating to sharing data on the web and concludes that the fundamental problem of "not wanting to share the toys in the sandbox" is the biggest impediment. With regard to technical issues, Buttenfield rightly draws attention to the problems of transmission capacity, which have led to the information superhighway being described as the "Information B-Road". However, Buttenfield's conclusion, suggesting a lack of efficient data compression algorithms, is very much focussed on the use of the Web to provide access to a digital library of spatial data, including vector and raster map layers, satellite imagery and scanned aerial photographs. I would contend that the Web will become less about the provision of spatial data and much more about the provision of geographical services and information. The information returned from these services will usually be rather less voluminous than sending the raw data. Most users of the web will not have the capability, or desire, to process raw data; their questions will be simple, but inherently spatial, for example "where is my nearest cash-machine?" or "suggest a tourist route around Scotland which will satisfy the diverse interests of my family?" The provision of data will become a specialist niche, focussed around public and private-sector intranets, which will furnish the raw materials for these service and information providers.

Because the data does form the basis for this information and service provision, it is important to ensure easy interchange of compatible datasets, and that requires a level of standardisation and organisation beyond that available at present. An effective infrastructure for the provision of data, to those who need it, is a very necessary first step. Thus, the data infrastructure forms a second theme running through this book.

The book is composed of twenty chapters, divided into five parts, which build on each other. First, we will consider the implications of a data infrastructure and the components which need to be put in place to realise the vision. The Web figures significantly in this discussion, and leads directly to the second part of the book, which brings GIS together with the Internet, both in terms of supporting the dissemination of information and as an analysis tool for the Internet itself. Spatial analysis rightly features within most GIS texts, and next we review various operational procedures, which I regard as being important, because of their immediate real-world application in terms of providing the sorts of services which will be based of the infrastructures we have set in place in Parts 1 and 2. Next multi-dimensional applications are reviewed, including the application of parallel processing, a key technology, necessary for high-performance data serving, which is quite critical in the Web environment. The fifth and final part includes some key spatio-temporal modelling applications, with the necessity of information infrastructures and the benefits of the Web being re-iterated.

1.2 CREATING A DATA INFRASTRUCTURE

Data infrastructures are essential as GIS becomes more closely integrated with broader Information Technology (IT), and the interchange of information becomes more critical. Whereas in the past, it could be argued that GIS has regarded itself as an island, this is ultimately an untenable situation. The need to share information, and indeed encourage a developing marketplace based on geographical information, is becoming recognised internationally. Indeed, countries as diverse as Canada, Portugal and Malaysia are all in the process of implementing national spatial data infrastructures (Masser, 1998)

Nancy Tosta critically explores the history, development and successes of the National Spatial Data Infrastructure (NSDI) in the USA, which has acted as both the impetus and the model which has been variously applied, to a greater or lesser degree, elsewhere.

She concludes that the NSDI has raised the visibility of geospatial data and the benefits of data sharing. Importantly, successes have been achieved as the result of high-level political endorsement and availability of funding to encourage data partnerships.

At this point, it is worth contrasting the NSDI with its equivalent initiative in the United Kingdom (UK), the National Geospatial Data Framework (NGDF). In the UK, there has been little 'high-level political endorsement' and certainly no funding (in the US, federal government cannot mandate other levels of government to undertake any action without providing all of the necessary funding!). The 'three pillars' of the UK NGDF are collaboration, the use of standards and access to geospatial data (Rhind, 1997). Perhaps unsurprisingly, with a lack of resources, without the political backing, and with a complete lack of comprehension in government of the potential for a new economic sector based on value-added geographic information, the NGDF has yet to make any real impact in the UK. Despite the best efforts of the NGDF Advisory Council, working closely with the Association for Geographic Information (AGI), the NGDF is regarded as almost irrelevant by most of the non-governmental data producers and the academic research sector. Neither yet sees the potential benefits for their own organisations, and this is compounded by the lack of resources.

Burrough and Masser (1998) provide a helpful review the broader European perspective, and this too provides an interesting comparison with Tosta's chapter here. The key differences relate to the European desire to treat geographic information as a tradeable commodity, yet surely the value of GI increases only when it is used? Lessons need to be urgently learned from Tosta's work.

Turning from problems to solutions, the chapter by **Claudio de Souza Baptista** and **Zarine Kemp** together with the following chapter by **Anup Pradhan** and **Bruce Gittings** provide complementary solutions to the problem of spatial data discovery on the Web. While search engines such as Alta Vista and Web Crawler provide useful tools for general-purpose web searches, clearly the identification and cataloguing of spatial data provides its own special problems. De Souza Baptista and Kemp review what is required of spatial metadata and present a model for its object-based structuring. A prototype web-based implementation is presented, which uses a Java client to interrogate an object-relational metadatabase, implemented over an Informix core. Pradhan and Gittings use a more traditional relational approach (using Oracle) to maintain the metadatabase, but extend the concept in two directions. First, a robot is presented, which can search the Internet (web and ftp sites) for spatial data. Second, server-based data conversion tools are demonstrated, which provide on-the-fly conversion, from the stored format, to that required by the user.

To close this first part, **Matt Duckham** and **Jane Drummond** discuss an object-based approach to data quality. This is an appropriate closing chapter to the first part, and follows naturally from what has been said before. Not only is it vital to understand and communicate data quality as part of any data infrastructure, but also at the pragmatic level of spatial data discovery, next to location, quality is perhaps the most important component of the metadata description. Duckham and Drummond explain how effective one strand of the U.S. NSDI has been, the National Committee for Digital Cartographic Data Standards (NCDCDS). They implement object model in Laserscan's Gothic system, to produce the core of an error-sensitive GIS.

1.3 INTEGRATING GIS WITH THE INTERNET

The growing importance of the web is emphasised once again by **Menno-Jan Kraak**, who explains the opportunities and problems the Web brings for cartography. The Web allows large numbers of maps to be published to an international audience at very low cost. However, Kraak warns that cartographic principles are ignored at our peril and new problems, such as the byte-size of the images, have to be handled otherwise the casual web browser will quickly lose patience and look elsewhere for their information. The diversity of techniques that can be used to portray graphical information on the Web are reviewed. Kraak returns to the theme of spatial data infrastructures by constructing an argument, which makes the case for web-based national atlases, as a by-product of the setting up of National Geospatial Data Clearinghouses. He sees such atlases as a national and educational resource, but also as a shop-window for the data providers products. A case study from the Netherlands is presented, which has been proposed by the Netherlands Cartographic Society, to their government, as part of their spatial infrastructure.

We have seen how the web can be used in the context of the standards and agreements, which form part of a spatial data infrastructure, for the provision of data. Now it is appropriate to build on this, and the delivery technology reviewed by Menno-Jan Kraak, to consider how *information* can be disseminated, and a particularly appropriate example is that of decision support.

Roger Dunham, **David Miller** and **Marianne Broadgate** show how a series of complex models, of the effect of wind damage on trees, can be made accessible to a user community through a web-based interface.   The models are shown in different configurations, including various forms of interface to GIS software.   The authors demonstrate the concept in an international context, with examples from United Kingdom, Portugal and Finland.   This allows the different problems which can result in these regions to be demonstrated, including the combination of wind with fire. I would regard this as a classic *intranet application*, where, with careful design, genuinely useful information can be generated from a spatial database, which is of benefit to front-line staff.   The authors appropriately raise the issue of legal liability relating to the outputs from the system, should anything beyond a demonstration be deployed in an internet environment.   By no means restricted to the realms of spatial information, this has the potential of becoming a significant problem, especially in an increasingly litigious world.

Adrian Moore, **Gerard Parr**, **Donna McCormick** and **Stephen McAlister** are also concerned with decision support, in a regional economic development context. Although their pilot project involves Northern Ireland, their proposal is for the implementation of an integrated system, strongly focussed towards local authority needs, across Europe. This raises a range of issues, including security, copyright and the costs of the underlying data, together with the pricing of the service provided.   The classic problems of comparability of datasets are brought into focus by this type of system. Lastly, the little-researched issue of the consistency of web-based user interfaces, particularly in relation to the use of icons, is reviewed. Various commercial web delivery solutions are used, which are based on combinations of client-based (Microsoft OLE/OCX and Java) and server-based techniques.

While we have thus-far considered the use of the Web, we should be aware that the Internet gives Geographers something they perhaps thought they would never have again, an entirely new space to explore, map and characterise. The Internet is a fascinating space, with many of the features of more traditional spaces, for example route-ways, communities, commercial activity, rapid population growth and, in some ways, highly structured. Yet, it has some rather curious effects, for example, although Washington is usually 'nearer' to Edinburgh than Paris in *internet space*, the very next milli-second it may not be, and the milli-second after that, it may have disappeared altogether! Researching this *internet space* gives us more than an abstract academic study; it has the potential to guide work that will address the very major performance issues that affect the networks. **Narushige Shiode** and **Martin Dodge** build on some of the recent work at University College, London, by helping us to explore and visualise this new space. Traditional tools, such as GIS and maps, prove useful in this exploration. The authors show that there are distinct patterns to the distribution of internet addresses, within the UK, and intend to broaden their research to incorporate Japan and the USA.

1.4 OPERATIONAL SPATIAL ANALYSIS

To provide the necessary information services on the Web, various spatial analytical tools will be required. There are distinct benefits from providing high-performance computers that can effectively undertake significant spatial analyses on huge datasets, and making these accessible through the Internet. Dedicated compute-servers, most probably based on parallel processing technology, will become a feature of internet information provision. Data can be analysed at source, with the rather-more-compact results transmitted onwards to the user issuing a query. **Stan Openshaw, Ian Turton, James Macgill** and **John Davy** take Openshaw's classic Geographical Analysis Machine (GAM), which is certainly an appropriately computer-intensive example, and make it available to the web. The GAM is a generic exploratory tool, which allows the examination of point data for evidence of patterns. Originally written in FORTRAN running on a Cray X-MP vector processor, the benefits of rewriting GAM in platform-independent Java are reviewed.

Another obvious candidate which will form the basis of internet information services is geodemographic analysis. Whether collected through visiting a web site, or using an affinity card in the supermarket, geodemographic information is actively being compiled into sizeable data warehouses, to be used for targeting consumers. It is therefore appropriate to include **Zhiqiang Feng** and **Robin Flowerdew's** work on improved methods for geodemographic targeting. The use of fuzzy methods, common in environmental work, but less often used in socio-economic applications of GIS, provides a means of broadening a target group beyond one demographic cluster. An example survey is carried out and although the benefits of the technique are not yet proven, the variation in results between traditional methods and the new method suggests further investigation is required and indicates potential benefits.

Ian Turton considers the concept of pattern recognition as a method for discovering trends in spatial data. Ever-increasing volumes of data are becoming available and this trend is likely to continue as national and international data infrastructures are implemented and new higher spatial and spectral resolution satellite platforms become available. This suggests that there is a need for more sophisticated methods which go beyond a simple (statistical) summary of the data. Turton proposes image processing and computer vision methods to recognise patterns and assist in the interpretation of this mélange of data. In particular, the use of these methods, for the identification of urban areas in Britain, is demonstrated.

1.5 BEYOND TWO DIMENSIONS

The need for multi-dimensional representations within GIS is increasing, yet I would still regard current GIS as lacking in this regard and therefore developments worthy of investigation.

The work of **Jasmee Jaafar** and **Gary Priestnall** follows naturally from that of Turton. Again, the problem is extracting useful information from potentially large volumes of data. Jafaar and Priestnall are concerned with automated building height extraction for land use change detection. Integrating information already available within a GIS as an aid to automated change detection has been shown to be beneficial and the addition of building height was regarded as important in refining this process. The methodology essentially involves the subtraction of a reference elevation model from the surface model generated from photogrammetry. The authors examine accuracy as an issue crucial to success.

In terms of operational terrain analyses, intervisibility analysis is perhaps the most important, for example in the rapidly growing mobile telephony market. Transceiver location requires the intervisibility analysis of prodigious amounts of terrain data, from a variety of sources, especially in the urban environment where resolution must be high. This brings together the need for a robust data infrastructure, with the use of high performance computing for repeated modelling of a constantly changing situation. Although this is not the particular application highlighted by **Mark Dorey, Andrew Sparkes, David Kidner, Christopher Jones** and **Mark Ware**, their work is of importance across a range of potential applications. These authors propose strategies for the enhancement of terrain models, including the incorporation of buildings, vegetation and other landscape features. Virtual Reality Modelling Language (VRML), which is closely associated with web-based visualisation, is used to identify inconsistencies between our digital models and the real world. Raster DEM structures are compared with TINs and when landscape features are included in the TIN, the accuracy of line-of-sight (LOS) calculations is seen to improve markedly.

These themes are continued through the complementary chapter by **Philip Rallings, David Kidner** and **Andrew Ware**. This examines the implementation of a reverse-viewshed algorithm and again picks up the theme of parallel processing. Various authors have been predicting a central role for parallel processing in GIS for almost a decade (e.g. Gittings et. al., 1993), yet this has not happened. While integrated data storage models, within formal database management systems, such as Oracle, have taken advantage of parallel storage and query options, the GIS algorithms and spatial analyses remain essentially serial in character (Healey et. al., 1998). The reasons for this are two-fold; first advances in the speed of computer processors are (just) managing to keep pace with the increased sophistication of analyses and, secondly, because of the significant costs of re-implementing algorithms in commercial GIS. However fundamental limits will be reached, and the volumes of data awaiting analysis will increase markedly. A further factor is the user's expectation of an instantaneous response from web-based systems, including those which may involve significant data processing behind the scenes. While overcoming the network delays associated with the web are more difficult, to minimise the overall delay, it is vital to be able to complete processing as quickly as possible and hence the need for parallelism.

Thus what Rallings, Kidner and Ware present is a rare example of a parallel implementation. Their implementation is based on a network of 24 pentium-based PCs, connected by a standard ethernet network, which provide significant performance gains. This is an excellent example of a highly beneficial parallel architecture, where no expensive specialist server is required, but an existing resource is used co-operatively at a time of inactivity. We could envisage, for example, the overnight use of a corporate PC network to run complex models.

As the final chapter in this part, **Nathan Williams** examines the future challenges which must be faced in designing a four-dimensional virtual-reality GIS, another computationally intensive procedure. Williams proposes moving beyond a Virtual Reality based on VRML and the World-Wide Web, to much closer integration with GIS to address applications from environmental planning to archaeological modelling. The capabilities of a VR-GIS are proposed and a formal definition provided. Williams examines the issues involved in such integration, including the user interface as well as the system architecture and data integration. In terms of data integration, Williams argues for standards which bring together the requirements of the GIS and VR communities, which adds weight to the need for all-encompassing and forward-looking standards as part of our spatial data infrastructure.

1.6 INFORMING THROUGH MODELLING

The multi-dimensional theme is continued into the final part of the book, which considers the use of modelling to provide information for decision support. **Adam Etches, Christophe Claramunt, Andre Bargiela** and **Iisakki Kosonen** examine a marriage of temporal GIS and traffic simulation models with the intention of supporting the management of dynamic traffic flows. The authors point out that current integration is based on a loose-coupling, with separate user interfaces and data transfer through files. The authors propose the design for a temporal GIS (TGIS), with a common reference database, accessible to existing traffic simulation models. The TGIS is implemented within an object-oriented DBMS paradigm, and specifications are presented for each of the main concepts used. The dynamic components of the system, which may be measured or simulated, include individual cars, traffic queues and traffic signals.

Claire Jarvis, Neil Stuart, Richard Baker and **Derek Morgan** return to the implications of differing data handling strategies. They are concerned with the extention of point-based observations to make spatially continuous predictions. The authors question whether it is better to interpolate the inputs to, or outputs from, a dynamic environmental model. This is important because these strategies can determine both the efficiency and quality of the results, yet the issue has not been well researched. The authors use the example of the Colorado Beetle. Temperature is a critical input variable in terms of the development of this insect pest, which poses a significant threat to British agriculture. The authors show that inconsistencies result when the outputs from the model are interpolated to determine spatio-temporal patterns, as against interpolating the inputs. For other examples, this is not necessarily the case. The authors present detailed recommendations as to the best approach to use in different ecological scenarios.

The theme of inadequate representations for spatio-temporal modelling within current GIS is continued in the work of **Ross Purves, Jim Barton, William Mackaness** and **David Sugden**. The authors note that spatio-temporal modelling is a common procedure in the environmental and physical sciences, yet to combine environmental models with traditional GIS software often requires loosely-coupling, together with high level programming. They suggest that rather than extending traditional GIS software to incorporate all of the necessary facilities, developers should make their software more 'open' to allow integration into the environmental modeller's armoury. It is worth noting that these issues are being addressed, at least in theory, by the Open GIS Consortium in, for example, their *Pluggable Computing Model* (Buehler and McKee, 1998). However, the GIS vendors are a long way from implementation.

The particular application considered in this chapter is the modelling of snow drift. Drifting snow can hamper communications, give rise to avalanches and is an important factor in managing ski resorts where wind can redistribute large amounts of snow very rapidly.

The linking of a model, such as this, to a web-based hazard warning system, for example the Scottish Avalanche Information Service (SAIS, 1998) offers significant potential, realising the goal of real-time data analysis providing valuable information both to planners and the general public. This is an application for which web technology excels, potentially providing GIS with something it has never had before; instant accessibility to all. Such a model will be successful providing the data are available and the sophistication of the models and GIS are well hidden behind a good user interface.

The system in the final chapter of the book, described by **Carsten Peter** and **Neil Stuart,** is another that could make use of the web as a dissemination medium, in this case for warning of a flooding hazard. Again, this chapter takes spatio-temporal process modelling as its central theme. These authors also consider the difficulties of linking models and GIS, and suggest these arise not only from the limitations of GIS, but also problems because age of the models, often inheriting code written for an earlier generation of computing equipment. The importance of the diversity of environmental data that is becoming available is emphasised and the chapter recognises the need for pre-processing the data, an aspect with which the data infrastructures we have argued for will undoubtedly help.

In contrast to the work of Purves et. al. in the preceding chapter, Peter and Stuart argue that simple spatial process modelling can be undertaken by extending the data structures and functions of existing GIS. Their example models the inundation of a low-lying flood plain area in the event of a river bank failure during a surge tide.

1.7 CONCLUSIONS

In this chapter, I have attempted to illustrate the themes that link the chapters in the remainder of this volume. I have argued the need for, and the responsibilities of, data infrastructures, which are being, or need to be, created to support GI applications across the public and private sectors, including academic research. These responsibilities include, but are not limited to, standards, quality assurance, organisational and governmental processes and an accessible dissemination medium. The prime candidate for a dissemination medium is the ubiquitous World-Wide Web, which has the ability to extend beyond simple data provision, and act as an information-access tool, usable by all sectors of society. Information generated from a variety of forms of spatial analysis, which can be integrated with traditional GIS and sophisticated spatio-temporal models, will provide visualisations from simple maps to complex virtual worlds in support of a range of applications, from the decisions of planners to providing real-time traffic predictions. The strategies for integrating models with GIS, to gain the benefits of the GIS data structures and supporting functionality, clearly remains the subject of debate and more work is required in this area.

High-performance applications servers available to both an organisation's intranet and the Internet will be required. These will most probably use parallel processing technology, which will be able to support the complex environmental and socio-economic models, based on an underlying spatial data infrastructure. In support this data warehouse, the days of the centralised compute server have certainly returned, but these are now several orders of magnitude greater in size than the mainframe we used to chastise. The gurus who evangelised over distributing everything to the desk-top have clearly got it wrong.

In bringing all of this to fruition, there is much more research to do, and exciting times ahead. By the end of this book, I hope that you will agree.

REFERENCES

Buehler, K. and McKee, L., 1998, The OpenGIS Guide: Introduction to Interoperable Geoprocessing and the OpenGIS Specification. The OpenGIS Consortium. http://www.opengis.org/techno/guide.htm

Burrough, P. and Masser, I. (eds), 1998, *European Geographic Information Infrastructures*. (London: Taylor and Francis).

Buttenfield, B. P., 1997, 'Why don't we do it on the Web?' Distributing geographic information via the Internet. Guest editorial in *Transactions in GIS* Vol. 2(1) pp. 3-5.

Gittings, B. M., Sloan, T. M., Healey, R. G., Dowers, S. and Waugh, T. C., 1993, Meeting expectations: a review of GIS performance issues. In *Geographical Information Handling – Research and Applications*, Mather, P.M. (ed) (Chichester: John Wiley and Sons), pp. 33-45.

Healey, R., Dowers, S., Gittings, B. and Mineter, M. (eds), 1998, *Parallel Processing Algorithms for GIS*. (London: Taylor and Francis).

Masser, I., 1998, An international overview of geospatial information infrastructures: lessons to be learnt for the NGDF. *NGDF White Paper*. Version 1, 1/7/98. http://www.ngdf.org.uk/whitepapers/mass7.98.htm

Rhind, D. W., 1997, Overview of the National Geospatial Data Framework. *Presented at the AGI – GIS'97 Conference*, October 1997 and *NGDF White Paper*. http://www.ngdf.org.uk/whitepapers/dgagi97.htm

SAIS, 1998, The Scottish Avalanche Information Service. http://www.sais.org.uk/

PART I

Creating a Data Infrastructure

2

NSDI was supposed to be a verb

A personal perspective on progress in the evolution of the U.S. National Spatial Data Infrastructure

Nancy Tosta

2.1 INTRODUCTION AND BACKGROUND

This chapter provides a brief background on the history of the United States National Spatial Data Infrastructure (U.S. NSDI), including the institutional and political climate within which it has developed and progressed to date. There is an increasing level of awareness about the NSDI within the U.S. geospatial data community, but exactly what progress has been made in promoting data sharing is difficult to measure. The views expressed herein are solely those of the author and do not represent the perspective of any past or current employer.

The relatively easy and inexpensive availability of federal government data in the U.S. has its origins in the formation of the democracy. Thomas Jefferson, the second President of the U.S. (1801-1808) and author of the U.S. Declaration of Independence wrote in a letter to James Madison (the third U.S. President, 1809-1817) in 1787: "And say, finally, whether peace is best preserved by giving energy to the government or information to the people. This last is the most certain and the most legitimate engine of government. Educate and inform the whole mass of the people. Enable them to see that it is their interest to preserve peace and order, and they will preserve them." Jefferson's writings strongly support his view that an informed public is essential to the workings of a democratic nation. He believed that it was the government's responsibility to make all information about the activities of government available to the people such that they might correct any perceived wrongs of government. This includes information about government decisions and policies, as well as scientific, economic, and geographic information.

The public's right to know and their right of access to information is reinforced with the belief that the public has paid, through taxes, for the collection of government information. The U.S. President's Office of Management and Budget (OMB) has reinforced the beliefs of Jefferson by issuing a Circular that establishes federal policy mandating that federal agencies provide access to their information (OMB, 1993). The following are direct quotes from the Circular:

> "Government information is a valuable national resource. It provides the public with knowledge of their government, society, and economy – past, present, and future. It is a means to ensure the accountability of government, to manage the government's operations, to maintain the healthy performance of the economy, and is itself a commodity in the marketplace."

> "Because the public disclosure of government information is essential to the operation of a democracy, the management of Federal information resources should protect the public's right of access to government information."

> "The nation can benefit from government information disseminated both by Federal agencies and by diverse nonfederal parties, including State and local government agencies, educational and other not-for-profit institutions, and for-profit organisations."

The Circular goes on to require that all federal agencies provide access to their data at no more than the cost of duplication, unless specifically exempted by laws or other policies. Federal laws also limit the copyright that federal government agencies can impose on data. Most federal geographic data are not copyrighted. U.S. federal government policies encourage the private sector, as well as state and local governments, academics, and not-for-profit organisations to re-package and re-use federal data, particularly if in the process the data become more usable to citizens. Although some may argue that private companies should not benefit at government expense, the jobs created and taxes paid from the private use of data are ultimately of benefit to the nation. With few exceptions there are no proprietary interests in federal data. Anyone (private or public) may add to, enhance, integrate, or pare down data sets as they see fit. This is all in the interest of making data more accessible to the public.

It was under this institutional umbrella that the National Spatial Data Infrastructure (NSDI) effort began in the U.S. Perhaps it was even because of this policy environment, that a situation existed where thousands of organisations were in the business of producing and/or using geospatial data. This meant that many agencies were duplicating each others' efforts either by copying (and perhaps modifying) data sets, digitizing paper maps to create digital files, and building data sets of greater (or sometimes lesser) resolution over geographies where other data already existed. Nearly every new GIS installation at any level of government meant that new data would be created, often to standards that were established internal to the organisation. This situation continues today.

In 1993, the President's Office of Management and Budget estimated that federal agencies were spending approximately $4 billion annually on the collection and management of geospatial data. Undocumented estimates were that state and local governments were spending even more. This situation continues to exist and with the proliferation of GIS, expenditures are likely even higher. Data sets were being duplicated, new data were hard to find, the quality of data varied dramatically, and the ability to share, integrate, and use data from other organisations was limited. The situation was 'ripe' for organisation and the federal government stepped in to help by endorsing the concept of a NSDI.

Discussions about the NSDI in the U.S. began in the late 1980's, primarily in the academic community, although federal agencies had also begun to use the term by 1989 (National Research Council, 1993). The Federal Geographic Data Committee (FGDC) was formed in 1990 by the Office of Management and Budget to help coordinate federal geospatial data activities (OMB, 1990). The committee was authorised to coordinate specific themes of geospatial data that were being created by a multitude of federal agencies, such as transportation data sets, vegetative cover, geodetic control, and hydrographic delineations. However, as a component of its responsibilities, the FGDC was also charged with developing a 'national digital spatial information resource' in cooperation with state and local governments, academia, non-profits, and the private sector. When the National Academy of Sciences began to use the phrase 'national spatial data infrastructure' to describe geospatial data coordination activities, the FGDC agreed to adopt the same phrase to describe national coordination activities to minimise confusion about national objectives.

By late 1992, the FGDC had evolved into a series of subcommittees and working groups. In early 1993, when Bill Clinton and Al Gore assumed the offices of President and Vice President, the level of interest in science, information technology, and cooperative partnerships with state and local governments was significantly enhanced. Meanwhile, the FGDC developed a strategic plan for the NSDI. Gore initiated a program to 're-invent' the federal government, primarily by increasingly devolving more responsibilities to state and local governments. The FGDC successfully encouraged inclusion of the NSDI into Gore's re-invention program and Gore acknowledged the importance of the NSDI in his National Performance Review Report in September 1993 (Gore, 1993). This support by Gore resulted in a significant increase in funding for the NSDI within the U.S. Geological Survey's (USGS) budget to support the FGDC. Expenditures for the FGDC and NSDI within USGS now total approximately $7.2 million annually and have been at this level since late 1993.

In the same time frame of the early 1990's, the Secretary of the Interior, Bruce Babbitt, recognising the relevance of geospatial data in addressing complex policy issues such as habitat protection for endangered species and the difficulties his own department faced attempting to integrate different agency data sets, decided to chair the FGDC. To have such a high level appointed official chairing a data committee was unprecedented. The recognition by Gore, the visibility of Babbitt, and Clinton's interest in successfully accomplishing the National Performance Review, led the White House to call for a Presidential Executive Order to implement the NSDI. This document was developed in late 1993 and signed in April 1994 (Clinton, 1994). Interestingly, despite all this seeming high-level political attention to the NSDI, for the most part the political objectives were driven by other motives (e.g. re-invention and endangered species). However, over time, both the Clinton Administration and Secretary Babbitt have come to be strong supporters of the value of coordinating and sharing geospatial data. In a fall 1998 speech, Vice President Gore mentioned support for GIS and the NSDI as one of several programs supporting community development (Gore, 1998). He also indicated that funding for the NSDI grant program would be increased in fiscal year 2000.

2.2 THE NATURE OF THE NSDI

The actions identified in the Presidential Executive Order were derived from the FGDC strategic plan. The primary objective of the NSDI was, and still is, to maximise opportunities for data sharing. Data sharing would help reduce duplication, make more data accessible to all, and contribute support to a variety of communities addressing common problems. Several specific program areas were defined in the 1993 strategic plan. These activities continue to dominate FGDC funding today. They are: Developing data standards; establishing means to find, access, and distribute data (referred to as the 'clearinghouse'); developing commonly needed data sets in consistent formats (framework); and, encouraging partnerships throughout all activities of the NSDI.

On one level, it has been easy to agree that data sharing and reduction of duplication in data are good goals. But the interpretations of solutions for how to accomplish these goals have been diverse and not always compatible. For example, a perspective of some federal agencies has been that data sharing means the expedited (e.g. more funding made available) production of data sets that were traditionally (and currently) within their purview. The goal of the NSDI effort in many federal minds, is the creation of nationally consistent 'layers' of geospatial data. Local or state agencies that might be expected to help share in the costs of this federal work sometimes held the perspective that rather than funding consistent small scale federal data development everywhere, a more appropriate effort would be to improve the resolution and timeliness of data in certain geographies of interest, perhaps where population is denser or environmental and other issues more pressing. For some organisations and individuals, the NSDI represented the opportunity to create nationally consistent data sets, for others it meant the right data at the right time developed through shared cost arrangements. One's interpretation of NSDI - whether it means the same data everywhere, the best data where needed, one data set for each geography, or multi-integrated and generalised data - is likely to affect one's course of action in evolving the NSDI. Hence, the reason for the on-going debate.

Besides the nature and content of the NSDI, the process for creating or evolving 'it' continues to be a much discussed issue, in fact, even more so than the content. NSDI 'management' options range from centralised, top-down, federal control for building a national 'something' to creation of a pseudo-government, non-profit representative national council for building a national 'something' to thousands of individual data sharing activities loosely connected with standards and common objectives that might collectively be aggregated for use over larger areas ('bottoms-up approach'). One point of view argues that someone has to 'be in charge' to make an effort such as the NSDI come to pass. Another view is that NSDI is so broad and involves so many organisations and potential data applications, that no one can possibly 'be in charge', but that many individual actions can facilitate more data sharing. The type of action needed to evolve the NSDI has been likened to the Internet, entrepreneurial and distributed. Some consider the federal leadership to be narrow and have called for the creation of a broader 'council'. Much of the funding to support the NSDI over the past six years has been federal (predominantly USGS budget). No other organisation has indicated a willingness to pay for much, if any, of the NSDI evolution. However, thousands of organisations continue to invest in GIS technologies and data. Whether there is a need to somehow 'represent' these interests in a hierarchical national council continues to be debated. This debate is often framed as 'who has a seat at the table' to make 'decisions' on the NSDI and countered with an observation that a large enough table does not exist for everyone to have a seat. Debates about the size of the table often seem to divert energy and resources from discussions on how to facilitate data sharing.

2.3 NSDI PROGRESS

Given the differences in points of view on the nature of the NSDI, measuring progress poses some difficulties. If one judges progress on the number of times NSDI is now mentioned in the GIS literature, then yes, much progress has occurred. After the Executive Order was issued, the visibility of the NSDI spread to geospatial data advocates in many other parts of the world, and many other nations have now also initiated SDI efforts of one form or another. The acronym NSDI has become a term of art, discussed and recognised throughout the geospatial data community in the U.S. in all its myriad interpretations. Numerous committees have been formed and hundreds of meetings have now occurred where NSDI was at least one of the topics of discussion. Is this progress?

If the definition of NSDI is nationally consistent data sets, then the effort has largely been a failure. Likewise, if one considers reduction in data duplication to be the goal, then NSDI appears to have had little success. However, there does appear to be more data sharing occurring. A question could be raised about the validity of a goal that seeks to minimise duplication in data sets. The original concepts in the first NSDI strategic plan envisioned that national data would be built through the integration and generalisation of local data sets. There seemed to be imminent practical sense in investing in data at the highest resolution affordable and then building on it for other lower resolutions of data. However, the primary difficulty in this thinking is that from both an institutional and data manipulation perspective, coordination is often more difficult and costly than simply recollecting the data. Additionally, the spread of GIS to more organisations addressing different issues creates a need for more data. Expecting that fewer data sets will satisfy all needs is unrealistic. Even in situations where the same geography is at stake, and similar resolutions of data desired, institutional interactions and mandates for slightly different data often make the effort of 'sharing' data more complicated and difficult than simply duplicating, adding nuances, and maintaining separate, though similar data files, which may then be shared.

Another way to examine progress, is to examine each of the NSDI activity areas supported by the Federal Geographic Data Committee, including standards, clearinghouse, framework, and partnerships. Examples of all of these activities can be found on the FGDC web site (FGDC, 1998a). Many of projects that have been initiated under the auspices of the NSDI are listed on this site, including those funded (at least in part) by the FGDC. The challenge in trying to understand whether they represent progress is that many of the details about these projects are lacking. The recipients of competitive grants funded by the FGDC are required to produce final reports analysing the results, but relatively few of these are listed as being accessible on the Web. Examples of some of the results for those projects with final reports are included in the descriptions of activities below.

The development of standards has been extremely slow. As anyone who has ever tackled standards development will acknowledge, the process requires inordinate attention to details, haggling over arcane and seemingly obscure points, and endless review. And once standards are developed, how to encourage and 'enforce' use are additional issues. A goal of promulgating standards that will be similarly used by all collectors of data is unrealistic.

However, some progress has been made. One of the first tasks undertaken by the FGDC was development of a metadata standard. 'Data about data' were perceived to be useful to data sharing efforts. The metadata standard was worked through a variety of groups, including representatives of the library community who are familiar with other systems that index, catalog, and describe collections. The FGDC formally adopted the standard in 1994, and variations on it have spread world-wide. Several of the reports on the FGDC web page describe metadata efforts. As an example, several organisations in the San Francisco Bay, California area appear to have been successful in cooperatively documenting their data with the FGDC standard and then providing access to the data via the Internet (BAGIS, 1998). Other organisations, however, indicate some frustration with the standard and limited acceptance of its use (see Wisconsin Land Information Board (1995) for a discussion about the state of Wisconsin's efforts or Gifford (1998) for a description of Montana state activities). While a globally consistent metadata standard may prove elusive, there does seem to be increased awareness of the value of metadata. The private sector has provided some support for this standard by creating tools that assist in the development of metadata.

A few other data specific standards such as cadastral, wetlands, and vegetation classification have also been developed, but their application has been limited. In general, they are not perceived to be of much use to many data collectors, particularly at the local level. One of the many challenges of standards development is the determination of exactly what to standardise. In many cases, the federal leadership of the FGDC subcommittees, charged with developing standards for various themes of data, has focused on standards applicable to federal data sets, but not necessarily on standards that relate to the most commonly collected data such as transportation. Many organisations when initiating a GIS effort that requires data development, will look to the federal government for guidance in standards, but given that few useful data standards exist, more often than not locals will simply adopt whatever standard is convenient, including one created internally at the immediate time of need.

The clearinghouse activity has now resulted in more than 85 sites that provide FGDC-compliant metadata. This is more geospatial data fully searchable on the Internet than ever before, but considering the phenomenal growth in the World Wide Web and use of the Internet, and the large number of world-wide GIS users, the number is quite small. The level of effort to develop metadata compliant with the FGDC standard is significant and most organisations have difficulty rationalising the value. Hundreds of workshops about metadata and the clearinghouse have been conducted, and much data is being served on the Web in the U.S., but only a small fraction of the data serving is being done according to FGDC standards for the NSDI.

The framework effort to date has also been far more discussion than action. The concept requires a great deal of coordination among levels of government and data producers. This kind of coordination has not been common in U.S. mapping efforts. Again, as with standards development, the need to agree on common approaches to represent data that can be used from local to federal applications is a major challenge. Many GIS users, who inadvertently become geospatial data producers, have little in-depth knowledge of data structures. Many federal agencies have little understanding of how their data are actually used. Large investments in proprietary software may limit the willingness to adopt more generic data representations. Coordination for framework data development is most likely to occur when all or several players are in need of new data, either as part of an initial GIS installation or during a data base 'upgrade'. These are the most likely times when agencies will consider adopting a new 'standardised' data structure. In the meantime, 'interested' discussion dominates over action.

The FGDC has been more successful in the arena of NSDI partnerships, partly perhaps because they have been willing to offer funding as an incentive to partner. The

Competitive Cooperative Agreements Program (CCAP) has funded several dozen projects that require an organisation to develop agreements or working relationships with other organisations to document data, share data, or in some way make data more accessible. Some of these efforts have fostered dialog and on-going commitments to share data, while others seem to have resulted in one or a series of meetings, with little data sharing outcome. One of the challenges in working relationships with state and local governments is that not all of them follow the federal policy of freedom of access to government information. In many instances, cost recovery for data production, fees and licenses for data use, and limited re-use policies have been adopted. These policies vary significantly among jurisdictions and states, and one of the major challenges in developing data sharing agreements is addressing these differences.

The FGDC has built partnerships with State governments by encouraging the formation of geographic information councils to coordinate interests in states, including federal, local, and private. To date approximately 27 councils have been recognised by the FGDC and are invited to participate once or twice a year in FGDC meetings in Washington, D.C. Travel is paid for by the FGDC. In some cases these state councils have been effective in organising statewide interests, in others, they touch only a very small fraction of geospatial data users and producers. The success of these organisations is often a function of political support (e.g. whether high level elected officials endorse the effort) and/or the presence of an energetic and committed leader.

The NSDI effort seemingly has improved relationships among federal and state agencies, particularly in producing data to federal standards (e.g. USGS digital quadrangles). However, in general, states are not big producers of new data, they are primarily re-users of existing federal data, which they may enhance or update. The primary data producers in most geographies are increasingly local government agencies, which tend to collect much higher resolution data than are used by state or federal agencies. The coordination between state and local agencies or federal and local in geospatial data management activities is nearly non-existent. Federal government agencies, particularly those that control operations from Washington, D.C., tend to understand how to make maps and build data sets that represent space, but have little understanding or appreciation for specific places. Local governments have deep understanding of places, but less comprehension about how those places should be measured in space.

One partnership interaction that seems to be growing is between local jurisdictions and private companies. For example, a local jurisdiction may contract to have data developed such as digital imagery or vector delineations from imagery, and will receive a discount in exchange for allowing the private vendor to subsequently market the data to other users. In these cases, however, jurisdictions and the companies they work with, consider the data proprietary, and only shareable at a price. (For a discussion of a practice similar to this, see the FGDC funded Dallas/Fort Worth Texas Clearinghouse Node project (Briggs and Waddell (1997).)

Most of the work that has gone into the NSDI as of this writing has been directly funded or underwritten by the federal government. The exact expenditures are not possible to ascertain, because many federal agencies do not label their geospatial data activities as NSDI. Roughly $35 million has likely been expended specifically on NSDI efforts to date by the USGS alone, in support of the FGDC and its work. These expenditures have included incentive grants for partnerships, a variety of research contracts, and travel funding for hundreds of state, local, non-profit, and university personnel to travel to various FGDC sponsored meetings. Some of these resources have been matched by non-federal participants, but the percentage is low.

In summarising progress, the NSDI has served to raise the visibility of geospatial data and the potential need for and value of data sharing. This has been the result primarily of high-level political endorsement and availability of funding to encourage data partnerships. In some situations, the concept or 'presence' of the NSDI has been used to leverage funds for cooperation. The NSDI has also contributed to the formation of many data coordination groups, including state GIS councils. However, the result of this appears in most instances to be more talk than action. In some situations, organisations have come together for at least short periods of time to share some of their data. The longevity of these relationships has not been determined, but simple increased awareness of other agency efforts likely goes a long way in contributing to data sharing. Has it been worth $35 million? The jury is still out.

2.4   A FUTURE FOR THE NSDI?

Based on the high-level political support for the NSDI in the U.S., the expanding discussions of the topic in other nations, as well as the continually growing base of GIS users world-wide and their insatiable appetites for geospatial data, it is unlikely that discussions about NSDI will terminate anytime soon. However, the technological and institutional changes that will both hinder and help evolution of the NSDI are likely to be dramatic. The networked, easy-access-to-Internet data world that exists today is very, very different from the world that prevailed in the 'early NSDI years'. The World Wide Web was virtually unknown when the FGDC conceived of a geospatial data clearinghouse in late 1992. Organisations such as the Open GIS Consortium, addressing issues of interoperability, were barely seeds of ideas. And the private sector was not a major marketer of digital geospatial data. All of these situations have now changed and are affecting how data sharing is conceived and implemented. Productive evolution of the NSDI must consider these real-world changes.

In the opinion of this author, the NSDI will evolve in several ways:

1.   Increasing efforts will be made to tie its development to real problems, such as Vice President Gore discussed in his recent speech - how do we create more livable communities? Other questions might deal with how to best respond to (or prevent) floods, or hurricanes, or how to protect endangered species habitat and national parks. The need to create multi-participant and regional organisations to address these types of issues creates a growing need to organise data across disciplines and organisations, which then brings thinking back to some form of an 'SDI'. This seems to be a healthy evolution and will continue to create more local interactions for data sharing.

2.   NSDI discussions will continue to spread throughout other nations, partly because of identified needs for data, standards, and coordination, but also in an effort to compete with the U.S. in the geospatial data field. Every country will, of necessity, approach the task differently. Different geographies, such as size and environmental complexity; diverse forms of government, including different relationships among levels of government; varying government policies, in particular, data access and intellectual property rights; and different issues being addressed by GIS; will all lead to a multitude of flavors of 'NSDI', not unlike what already exist in the U.S. In some regions, discussions among nations are taking place, such as European

discussions about a European Geographic Infrastructure, under the auspices of the European Commission. As these happen concurrently with national efforts, tensions build around the questions of 'who's in charge?' and 'who is advising whom?' It's the top-down versus bottom-up question again.

3.   While recognising the energy being expended on national efforts and preliminary discussions on a global spatial data infrastructure (GSDI), the successful future SDI's will be local in nature. This is as much a function of practical matters such as the challenges of coordinating large numbers of people over large areas, as it is recognising that all geography is local and issues, physical characteristics, and institutions vary significantly across nations and the world. In some ways, perhaps it is the 'N' in NSDI (or 'G' in GSDI) that creates debate, if only because of the complexity of geography, interests, and disciplines that must be engaged in most nations. The only likely successful national efforts may occur in very small and less complex countries, such as Qatar (Tosta, 1997).

4.   The FGDC framework concept of building state or national data sets from the ground up based on local data will be abandoned. The concept that data covering a large geography can be 'knitted' together by horizontally integrating many consistent local data sets is worthwhile, and in some cases may be implemented. The practice is more likely to occur if specific issues must be addressed. The USGS undertook such an effort after massive flooding in the Mississippi River. Hundreds of databases from numerous sources were integrated and then stored on USGS computers at EROS Data Center in Sioux Falls, South Dakota. However, today, no one maintains the data base. For the most part, local data are not consistent across jurisdictions. The diversity of geospatial data producers, their various stages of production, and differences in the sophistication and application in use of GIS make the goals of consistency and easy integration elusive ones. What are analysts who require data over large areas to do? Most likely, given the processing power that now exists, organisations interested in issues that cover large geographies - nations and worlds - may find the means to simply create consistent, small scale, large-area data from raw source material. An example of this might be development of a simple land use data base from satellite imagery. One factor that could conceivably change this state of affairs is that enough funding be made available to provide incentives for locals to create consistent data sets. The expense would be enormous.

5.   Although the FGDC metadata standard in it's entirety is not likely to ever earn a large loyal following, the recognition of the value of metadata will continue to grow. As more data are Internet-served, the requirements that those data be searchable in a variety of ways will increase. Documentation of data will become an accepted expenditure, along with increased emphasis on data maintenance.

6.   The NSDI will never be a tightly organised activity. No one will ever control, coordinate, or be in charge of the entire effort. Attempts to try to develop a representative council or bureaucracy to manage the NSDI will continue to divert energy from real discussion about spatial data sharing and technology issues. The likelihood that an organisation could be created that would have as much clout in terms of influencing data producers, as the little currently wielded by the FGDC, is very low. While the federal government and it's partners may continue to 'mandate' various activities and standards, most organisations will apply for and accept the limited FGDC funding, make an effort to use national standards to the extent feasible, and then do what needs to be done to meet their immediate needs. This situation would not be any different should a 'pseudo-government' organisation 'be

in charge'. The diversity of needs, interests, and players is too great. Immediate needs always supercede grand agendas. Given the current political support for the FGDC, the most constructive near-term action would be for interest groups to lobby, engage, and make use of the committee to address their issues and concerns. The existence of such a highly visible political body, that meets solely for the sake of discussing geospatial data, and a support staff with funding to spare, are not opportunities that should be ignored by anyone with interests in geospatial data.

7.   The future will likely see an increasing presence of the private sector in these data discussions. Their involvement could be an impediment to data sharing (most companies would prefer to sell a product several times over to a group of clients, rather than have them all share 'one copy') or an incentive, as they may provide funding to specific groups in exchange for data distribution rights as discussed previously. Data and information have become too much of a 'currency' to be ignored by the private sector. One potential implication of private sector involvement and a trend that dominates most of the world, and appears to be gaining favor in the U.S., is the commodification of data. More agencies appear willing to consider data cost recovery policies, than has been true in past U.S. history. This is certainly contrary to Jeffersonian philosophy and current federal data policies, but represents a practical reality as demand for data increases. As public agencies act more like private companies, this trend will continue.

2.5 CONCLUSIONS

The national program of the NSDI within the U.S. evolved out of a situation of large numbers of GIS users creating geospatial data sets and a federal policy encouraging access to information. The effort has seen some success, primarily in raising the awareness of the value of data sharing and data documentation, but also in some specific instances where groups have come together to share data over a common geography. However, the program has cost at least $35 million to date, with more funding promised. The ability to judge the return on this investment is difficult due to diverse perspectives on what the NSDI is for and how it should be managed. The criteria to measure progress or success have not been agreed upon. Nor has there been an effort to publicaly re-examine the assumptions originally made in the 1993 FGDC Strategic Plan about the activities which might best serve the evolution of the NSDI.

The question of whether the objectives and activities of the NSDI make sense over time, must be continuously asked. The fact that the FGDC was created and the concept of the NSDI adopted nearly concurrent with a political administration that supported technology, was partly luck in timing, but mostly a readiness with ideas that were based on current technology and political objectives, not past programs. The networked, tuned-in world of the late 1990's is very different than the isolated, incompatible-data world of even the early 1990's, when the FGDC and NSDI were in their first throes of organisation. There are constantly more users of GIS, more data, more applications, more hardware, more software, and more complex institutional interactions. The requirements for data change constantly as the issues, players, and technology change. The ability to create standards in this state of flux is likely to be extremely difficult, if not impossible, and the expectation that a few shared data sets will meet most needs is naive. The on-going debates about the NSDI might better serve the public if the focus moved from 'who's in charge?' to 'what actions are needed to make better use of technology to share data?'.

The expectation that a constantly growing and changing 'beast', such as the geospatial data user community, can be managed and coordinated is not unlike expecting to coordinate the global economy. Certainly some institutions play key roles, but no one is in charge and no one can really explain how it works. Individuals, institutions, and countries react and respond to a constantly changing environment that is linked worldwide. The NSDI is not very different. NSDI was conceived to be a verb, recognising that action, interaction, and reaction are required to promote changes in behavior that might better support data sharing. The perception that NSDI was and is a passive thing that can be constructed by someone, misunderstands the objectives and may impede future opportunities for evolution.

Further, one should not underestimate the role that human nature plays in any endeavor based on interaction and sharing. Many individuals appear to prefer the role of 'coordinator' to being coordinated. People want recognition for their work, the ability to claim credit for work well done, and some control over their time. NSDI thinking originally recognised this behavior, which was the basis for suggesting a bottom-up approach. If the federal government attempted to tell state and local governments what to do with geospatial data, conversations on NSDI in the U.S. would have halted years ago. The small competitive grants oriented to local jurisdictions were intended to provide funding for building capacity to enter into data sharing arrangements, the outcome of which would be data sharing at all levels of government (see FGDC (1998b)). To some extent, as noted previously, these efforts have worked and represent some of the small successes of the NSDI. On the other hand, there will never be enough money to provide incentives for everyone to be encouraged to change their behavior, and the small amounts of money to each organisation will not necessarily have a long-lasting effect. The benefits and criteria of such a grant program may need to be rethought.

This diversity of perspectives on the NSDI in the U.S. is multiplied ten-fold in other nations of the world. The spread of the concept of NSDI to other nations will likely result in unique interpretations of goals and actions for development of spatial data infrastructures. This is already occurring. For the most part, other nations have more stringent copyright protection on national data sets and charge higher costs for accessing data, creating very different issues in the debate on development of an NSDI. The whole discussion about data sharing is colored by issues about price. These differences will be the issues that will be debated for many, many years in global spatial data infrastructure discussions.

In closing, the key strategies needed to evolve the NSDI further are summarised as follows. First, political opportunities should always be seized. Visibility is usually better than invisibility. Secondly, a primary goal should be established and it should be to improve access to data. This will lead to additional objectives as the questions about how to improve access are asked. Thirdly, in efforts to improve data access, bureaucracy and central control should be minimised. Fourthly, technology changes and accompanying organisational changes should be tracked and taken advantage of. Fifthly, evolution should be anticipated. Nothing will stay the same. The activities of yesterday are not likely to be the same as tomorrow's. And finally, patience must be practiced. The NSDI will only ever evolve piece by piece, edge by edge. No one and nothing will every coordinate the whole shebang. Small successes should be appreciated and failures should be debated, re-thought, and learned from. Jefferson would most likely be pleased at how his original wisdom has been played out.

REFERENCES

BAGIS, 1998, San Francisco Bay Area demonstration GIS project: Metadata for datasets relating to the BAGIS project, www.regis.berkeley.edu/bagis/metadata.

Briggs, R. and Waddell, P., 1997, Establishing a National Geospatial Data Clearinghouse Node for the Dallas/Fort Worth Vote Region: A model for metropolitan areas, www.brunton.utdallas.edu/research/usgs/usgsmeta.html.

Clinton, W.J., 1994, Co-ordinating Geographic Data Acquisition and Access: The National Spatial Data Infrastructure. Presidential Executive Order 12906, April 11. Washington, D.C.

FGDC, 1998a, Federal Geographic Data Committee, www.fgdc.gov.

FGDC, 1998b, National Spatial Data Infrastructure Funding, www.fgdc.gov/funding/funding.html.

Gifford, F., 1998, Montana GIS Data Clearinghouse Final Report. http://nris.msl.mt.gov/nsdi/finrpt96.html.

Gore, A., 1998, Remarks by Vice President Al Gore. Speech at the Brookings Institution. September 2.

Gore, A., 1993, From Red Tape to Results: Creating a Government that Works Better and Costs Less. Report of the National Performance Review. Washington, D.C.: U.S. Government Printing Office.

National Research Council, 1993, Toward a Coordinated Spatial Data Infrastructure for the Nation. Washington, D.C.: National Academy Press.

Tosta, N. 1997, Data Revelations in Qatar: Why the Same Standards Won't Work in the United States. In Geo Info Systems, Vol 7, No. 5, pp. 45-49.

U.S. Office of Management and Budget, 1990, Circular A-16: Co-ordination of Surveying, Mapping, and Related Spatial Data Activities. October 19. Washington, D.C.

U.S. Office of Management and Budget, 1993, Circular A-130: Management of Federal Information Resources. June 25. Washington, D.C.

Wisconsin Land Information Board, 1995, Wisconsin NSDI Clearing House Initiative Final Report, http://bager.state.wi.us/agencies/wlib/sco/pages/finlrpt.htm.

3

Spatial Information Systems and the World Wide Web

Cláudio de Souza Baptista and Zarine Kemp

3.1 INTRODUCTION

The opportunities for developing digital geospatial libraries and providing global access to such data have been fuelled by the ever-increasing use of the Internet as a communication medium. The requirements addressed by the World Wide Web (WWW; the terms 'the Internet' and the 'World Wide Web' are used interchangeably in this chapter) are many:

- generating and maintaining geospatial data sets is expensive and time-consuming so sharing the data as widely as possible is important;
- data collected at local, regional and international scales require to be collated depending on the spatial focus of decision making processes;
- geospatial data from diverse sources, satellite imaging, monitoring instruments, GPS receivers etc, need to be integrated for spatiotemporal analysis.

The case for geographic data dissemination and integration at the European level is made cogently in Burrough *et al.* (1996). The authors provide a typology of geographic information based on its relevance to various activity areas and identify the key issues involved in the provision of a European framework for geographic information. One of the issues that is raised repeatedly is that of metadata, the importance of which is discussed in greater depth in section 3.3. The terminology associated with the activities which can be described by the term 'GIS and the Internet', such as 'liberating', 'empowering', 'enabling', tends to be very positive. However, practical experience of users at the end of an Internet browser is less emphatically favourable. MIDAS (Manchester Information Data Sets and Associated Services) found that the uptake of spatial data sets was limited (Li *et al.*, 1996) and suggested several causes for this effect.

This chapter presents a discussion of user perspectives on access to geospatial data repositories via the WWW, focusing on the importance of metadata. It describes a hierarchical metadata model which provides different levels of abstraction, presents a system architecture for the underlying model and describes a Java-based prototype used to experiment with access to geospatial data via the Web. The chapter concludes with a summary of research issues.

3.2 PERSPECTIVES ON SPATIAL DATA ACCESS

The typology of user requirements for spatial data range from the search for information about a specific theme with reference to a specific part of the earth's surface to the acquisition of detailed spatial data sets for use in spatial analysis procedures. Consider the domain of fisheries and marine environments: a user might wish to know about the prevalence of the use of a certain type of fishing gear, for example, 'purse-seiners in the Mediterranean' or 'the level of discards off the Algarve coast'. These are fairly open-ended queries which can be responded to by the resource discovery services provided by web search engines such as AltaVista, Yahoo and others of that ilk. These search engines use either automatically generated textual concordances or human-mediated indexes to surf the global information base and provide pointers to the required web pages. Carefully structured boolean searches do a reasonably good job of eliminating ambiguity and providing a usable subset of URLs (Universal Resource Locators) that contains the required information. There is no requirement on the users' side for a specific data set and no guarantee on the part of the search engine to provide the best subset of URLs. The data retrieved in the form of HTML pages can be very effective, in multimedia form, constrained only by the creativity and imagination of the originator of the web pages. Thematic maps, and an extension of these, 'active maps', that enable minimal spatial interaction with the user are a very effective means of visualising and conveying spatially referenced information. The interaction with the user can be extended to take account of variations in scale; the volume of data and the thematic context being determined by the spatial scale requested by the user. There are several, extremely effective examples of this sort of spatial web-based searching; retrieve information about 'pollution incidents in a specified marine area' or 'within 10 km of the point identified on the screen map'. The retrieved web pages can also lead the user to browse through a set of related pages with contextually relevant information. This type of spatial search lends itself to serendipitous information discovery as long as there is no need to use the retrieved information for further spatiotemporal analysis.

At the other end of the scale, a user may wish to retrieve spatiotemporal information as part of ongoing research effort. For example, 'retrieve data on total catch of fish species within spatial area x_1, y_1, x_2, y_2 within time interval t_1, t_2'. In addition, data on marine environmental variables such as temperature, salinity, and planktonic abundance may also be required for the subset of space-time that is of interest. This type of query makes very different demands on web services and requires an effective searchable set of metadata to be provided.

3.3 METADATA FOR SPATIAL INFORMATION

There are problems with accessing geospatial data in the free-wheeling style of general purpose search engines; the indexes are not refined enough to cope with the detailed focused queries required for spatial analysis and these general-purpose systems cannot possibly provide them. To adequately index spatiotemporal data sets necessitates the use of metadata, informally described as 'data about data'; the information that describes the content, quality and other characteristics of the data. The GISDATA document on European geographic data describes a metadata service as "... probably the single most important action immediately needed" (Burrough *et al.*, 1996). The importance of metadata is reflected in the numerous schemas proposed and accepted that have evolved with the increase in web-based data retrieval. In the geospatial domain, metadata standards such as the Catalogue of Data Sources (CDS) of the European Environmental Agency and those of the Federal Geographic Data Committee Content Standard (FGDC, 1994) for the definition of the content standards and semantic definitions for geographic metadata are widely used and integrated into metadata schemas. A consequence of the complexity of geospatial indexing is that numerous variants of the basic metadata formats and schemas have evolved, to cope with the semantic complexity of various spatially referenced domains and the different data structures underlying them.

3.3.1 Metadata for resource discovery

When the use of spatial data repositories is considered from a global perspective, the problem of web-based access presents additional challenges. It is assumed at this level that geographic information users will cluster into communities with shared interests and perspectives as suggested by the OpenGIS Consortium (OGC) (OGC, 1998). Even with a shared conceptual core, problems arise in organisation and interoperability across heterogeneous multi-source, multi-platform, multi-standard computing environments (Lockemann *et al.*, 1997). Treating disparate, distributed spatial data sets in a unified manner implies providing:

- a high level conceptual metamodel that encompasses the individual data sets;
- transparent routing of queries to the appropriate server;
- a generic library of geospatial functions for co-ordinate transformations, format conversions, and standard interfaces (Clement *et al.*, 1997).

Here we concentrate on the first of these, the provision of metadata at a high level of abstraction to enable spatial resource discovery.

We note that there are several ways in which accesses at the global level can be organised. For example, one or more 'trusted sites' may maintain the equivalent of a catalogue service for all data providers within the community. This is the route taken by the communities for the Committee for Earth Observation Satellites (CEOS). Generic search and retrieval protocols such as Z39.50 (ANSI/NISO Z39.50, 1995) and the more specifically spatial protocols, CIP (Catalogue Interoperability Protocol) for satellite data and FGDC/GEO for geospatial data repositories enable disparate, distributed data holdings to be searched using a standard set of rules. These metadata models are heavily dependent on middleware software layers to maintain up to date catalogues of collections and route queries to appropriate databases (Best, 1997). A similar but slightly different approach is the provision of an enabling services system built on a set of linked middleware nodes which provide a conduit between data providers and end users (Kjeldsen, 1997).

It is obvious that the provision of open-ended global resource discovery capabilities is a major problem requiring considerable co-operation and contribution from participating sites. However, as a first step, smaller data source nodes can maintain metadata at the highest level of abstraction to provide the first stage in the resource discovery process. This high level metadata should be centred on the concept of a *collection*; a notion that has been adopted by the CEO as well OGC standards, as well as other digital library projects. The metadata hierarchy, at this level should consist of references to the underlying data sets in the form of descriptors or attribute name-value pairs that encapsulate the ontology of concepts used by a particular geographic community. We are of the opinion that metadata at this level should be directed towards providing a lightweight, high level indexing service rather than a detailed specification of underlying data sets. Initial searches guide users to a narrower number of interesting data sets which could be explored in greater detail to determine whether they should be downloaded for further analysis.

3.3.2 Levels of spatial metadata

The WWW has brought to the forefront issues of coping with vast volumes of heterogeneous spatial data in an efficient, integrated and interoperable way. This has not only highlighted the relevance of metadata, but also made the GIS community reflect on what is meant by metadata, its uses, scope and content, at different stages in the spatial resource discovery and data access process. At the core of the OGC standard, mentioned in the previous section, is the common geographic information component model. On the top of that is a hierarchical layer of metadata objects that relate to feature collections, their identification, and lexical, project and use semantics that enable sharing and integration of spatial data (Buehler, 1996). Research into provision of such web-based access to spatial data has led to the following conceptual architecture which identifies the metadata requirements at different levels in the spatial data dissemination process (Figure 3.1).

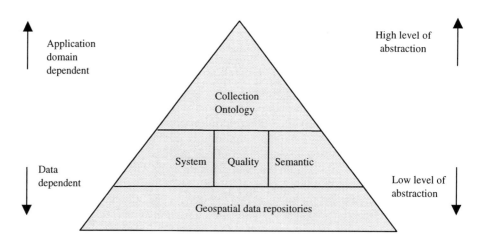

Figure 3.1 Metadata hierarchy

Consider a bottom up approach to the metadata objects identified. Broad categories of metadata include:

- *system metadata*: at the lowest level, this metadata is data type dependent and concentrates on the formats, protocols and vocabularies required to manipulate the objects. For example, for a satellite image, the metadata would contain information pertaining to the size of the image, the encoding mechanisms, the storage structures and so on. Metadata at this level is required to enable spatial data to be manipulated, transformed and displayed;
- *quality control metadata*: enables users to assess the fitness for purpose of a particular spatial object or set of objects. This metadata records information regarding the date, time, method of data capture, any pre-processing involved and maintains a history of the lineage of spatial data. Examples in this category include the algorithms used to process and transform an array of raw cell values into a classified raster object;
- *semantic metadata*: this category of metadata can take many forms and is designed to enable effective searching of spatial data. For example, the land cover classification hierarchy underlying the data set or the containment hierarchy of spatial objects may be captured as part of the associated metadata. In the case of image objects content-based attributes of the data may be extracted and stored for subsequent search and retrieval. Transducers may be used for automatic feature extraction thus providing a set of queryable attributes associated with a spatial object. Examples of such systems are described in: Ogle and Stonebraker (1995) which provides facilities for content-based analysis of image objects to enable 'concept queries'; and, Smith and Chang, (1996) which provides capabilities for image analysis for automated region extraction and joint content-based and spatial querying.

It should be noted that the metadata objects discussed so far have the following properties:

- the metadata objects may have a one-to-one or one-to-many relationship with individual spatial objects;
- the metadata objects themselves form a hierarchy and/or an aggregation of objects that collectively enable browsing the spatial object base in navigational mode as well as associative access to specific objects in response to structured queries;
- the metadata types discussed so far have been implicitly associated with one data provider i.e. with an integrated spatial database. By integrated we assume that although the database may consist of disparate spatial object types the database will be managed by one software package; either an object-relational database system with associated spatial functionality or an integrated GIS package.

This means that the extent to which that database is made available to users over the Internet and the interface provided, is under the control of the autonomous data provider. This is the premise under which the prototype discussed in section 3.4 operates. The client interface can be in point-and-click style or use a more formal spatial language. Moreover, the data provider may make selected data and functions available or may adopt an open-ended database query mechanism. In the latter case, the database schema is made available to the user who may query and retrieve data over the entire database using a structured query language such as SQL3 (Melton, 1995) with additional user defined functions for spatial analysis, or using a more intuitive graphical mode which is subsequently converted into a structured query language.

3.3.3 Metadata model design

In order to achieve the multiple levels of abstraction present in Figure 3.1, a model has been designed to encompass three levels: collection/ontology, semantic features, and meta objects. Figure 3.2 shows the classes and relationships of this metadata model.

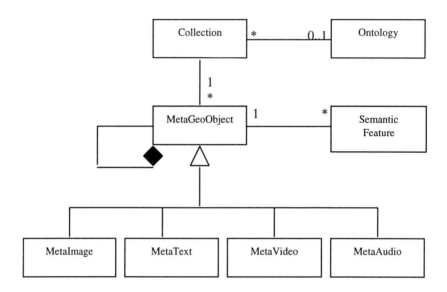

Figure 3.2 Metadata conceptual model

Collection: all represented objects are stored in the collection to which they pertain. The collection concept comes from the library domain. It is a structured way of organising objects in a hierarchical model. For example, a collection could comprise objects related to forest fires around the world, cities of England, or environmental disasters. One collection is related to many MetaGeoObjects that can be of different data types: satellite images, maps in vector format, free-text, video and audio. Typical attributes for a collection are:

colId: object identification for a collection;
colTitle: title for the collection, e.g. 'Environmental Accidents in Britain';
colDescription: a textual description of what that particular collection represents;
colResponsibility: agency or person who is responsible for generating and updating the collection;
colSpatial: overall spatial footprint of the collection;
colTimerange: temporal range of the whole collection, e.g. 1975..1994.

A collection can have associated with it an ontological structure which represents a searchable vocabulary related to the application domain.

Ontology: at this level, the system manipulates a vocabulary restricted to the collection domain. This vocabulary will be used for querying the geodata repositories at a high level of abstraction. An example of a classification is the European CORINE land-cover classification which proposes a three-level hierarchical classification. Using this ontology a user can pose a query to retrieve objects relating to 'coniferous forest'. If 'coniferous forest' is a concept present in the ontology, the system looks for the objects which represent this concept by using the pointers that associate that concept to the relevant object in the geospatial data repository. Figure 3.3 shows an example of a metadata instance at the ontological level. The ontology class is composed of the following attributes:

> *ontId*: the ontological object identification;
> *ontTitle*: the title for the ontology; and a set of tuples:
> > {<ontName, ontSynonym, set-of {<ontSpatial, ontTimerange ontDatatype>}>}
> > *ontName*: the feature name;
> > *ontSynonym*: links to related terms;
> > *ontSpatial*: the spatial footprint of the object;
> > *ontTime*: the temporal range;
> > *ontDatatype*: the data type e.g. image, vector map, text, video, audio or URL.

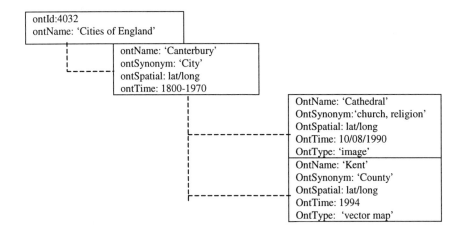

Figure 3.3 Ontology level metadata example

Semantic feature: at this level the system represents semantic knowledge which is extracted automatically from some of the underlying geospatial objects types in order to provide an efficient indexing mechanism. This level implements the semantic level depicted in figure 3.1; it contains the information necessary to realise content-based retrieval in both images and video geo-objects. The semantic features may, for example, contain colour and texture histograms. When the image is entered into the system, a metadata extractor generates the colour and texture histograms for that image and stores each of them in feature vectors in the respective semantic metadata class together with related information such as spatial footprint, temporal range, and data type (image or video). Using the metadata objects, for instance, users can pose queries to return images based on similarity measures extracted from the image e.g. 'select images similar(queryImage)', or 'select images where landCover('coniferous forest') > 70%'.

MetaGeoObject: at this level the system represents type specific metadata as well as system and quality control information. This class is composed of a hierarchy in which the inheritance is represented by an is-a relationship. At the root of this hierarchy is the geo-object class itself which contains attributes:

> *geoOId*: the object identifier;
> *geoName*: the object name;
> *geoAnnotation*: a textual description of the object;
> *geoCollector*: the person or entity who collected the data;
> *geoSpatial*: the spatial footprint;
> *geoTime*: the temporal range of the object;
> *geoHistory*: the history of the lineage of the geo-object;
> *geoCapture*: the capture method used;
> *geoPreprocessing*: describes any pre-processing algorithms applied to the data.

Four subclasses are derived from the MetaGeoObject class: MetaImage; MetaText; MetaVideo; and, MetaAudio. Those subclasses inherit the attributes and methods of the MetaGeoObject class and define their own ones. For instance, the MetaImage class has the following attributes: image type (e.g. remote sensing, aerial-photography, and photography), image size, resolution, format, and a pointer to the raw image. The class MetaText has the attributes: format (e.g. HTML, XML, ASCII), length, language, keywords, and URL attribute denoting an object which is not stored in the database itself but can be accessed at the URL address via the WWW. The MetaVideo class has the attributes: format, duration, and number of frames per second. The MetaAudio class has attributes coding, language, duration, number of channels.

An important concept in the MetaGeoObject class is the support for composite objects. In this case an object can be composed of several others forming a complex object. Therefore, it is possible, for example, to have a composite multimedia object which may consist of an image and an associated text object. The vector geo-objects (point, line, polygon) are instances of the MetaGeoObject class.

3.4 TECHNICAL ISSUES

In section 3.2, we noted that the user requirements for Internet-based spatial access could encompass a wide spectrum. Once the spatial data retrieval requirements transcend the basic textual keyword capabilities of the search engines and the stateless nature of the Hypertext Transfer Protocol (HTTP), most data providers move to a solution that involves using an object-oriented or relational database at the server end. This is the method that has been adopted in (Newton *et al.*, 1997) and by many other geodata providers. In some cases, the backend spatial data sets are stored in a proprietary GIS such as Arc/Info (*Li et al.* 1996). However the trend towards open, interoperable, distributed GIS militates against closed monolithic architectures and favours federated interconnected solutions. Developments in database technology such as object-relational and object-oriented databases provide the capabilities required for the underlying framework (Lockemann *et al.*, 1997). They support complex type structures and provide extensibility to enable abstract data types (ADTs) and user defined functions (UDFs) to be defined. These features are necessary for the specification of spatial objects in heterogeneous data repositories, the associated metadata at the collection and object level and the functionality required to index and manage the objects. In addition, they provide the traditional advantages of database management systems such as concurrency, recovery, query tools, persistence and distribution (Güting, 1994). The technical issues that arise in the design of underlying framework are concerned with connectivity of spatial databases, distribution of the processing load and scalability.

- Connectivity: technical solutions to provide connectivity between a database at the server end and a platform independent WWW browser at the client end include:
  - The use of web gateways via the Common Gateway Interface (CGI) which enables programs written in a variety of programming languages to add spatial retrieval services to the WWW;
  - Alternatively direct access between the client browser and the backend database can be provided by using a Java enabled browser running a Java applet through a JDBC (Java Database Connectivity) driver (Hamilton *et al.*, 1997). JDBC is an application programming interface (API) that allows a Java applet or program to connect directly to a database server using SQL commands;
  - Other solutions include using a JDBC-ODBC bridge (linking JDBC and Open Database Connectivity standards), the addition of the HTTP protocol into the DBMS functionality and plug-ins and proxy servers (Ehmayer *et al.*, 1997).
- Client-server load distribution: one of the main ideas behind the client-server architecture is to enable the distribution of the processing load between the server and the client. The main issue with this architecture is to determine the optimum split between the processing allocated to the server and that performed in the client. Depending upon the design decision taken, either 'fat' or 'thin' clients are produced (Franklin *et al.*, 1996). A fat client can cope with the Graphical User Interface (GUI) and also enables additional functions to be executed. For example, image rotation, zoom, pan, and even query processing can be executed at the client end. Fat clients also contribute towards reducing the communication traffic over the network by limiting the volume of interactions with the server. On the other hand, thin clients are lightweight; they are able to execute simple GUI functions but transfer all query execution to the server end.
- Scalability and performance: network bandwidth and server processing capability are the main concerns relating to the scalability of software operating through the WWW. Those issues can sometimes result in poor performance when the number of accesses tends to be large as in a spatial system which usually supports huge volumes of data (e.g. satellite images), involving terabytes of information. A more complex strategy may be required in order to deal with several concurrent clients with only a minor decrease in performance. For instance, network traffic may be reduced by using image compression and incremental queries (Andresen *et al.*, 1996; Bergman *et al.*, 1997). The former results in a decrease in the volume of data transmitted.

Incremental strategies transmit spatial objects in stages; instead of transmitting the complete object, say an image, on the initial request a thumbnail image or a sub-region of the image is transmitted. When the final decision is made, on a second request, the full image or the remainder of the image can be transmitted to the client.

3.4.1 Experimental prototype

In order to implement the hierarchical metadata model described in section 3.3, the system design was based on a three-tier architecture (see Figure 3.4). The client side is responsible for the GUI with spatial functionality to traverse and selectively query data. This interface is responsible for multimedia display (image, maps, video, audio, text) and interaction with the user. Access involves not only the data itself but also the metadata objects at all levels of the metadata hierarchy. The client is being developed in Java and the communication with the middleware server is made via Java remote method invocation (RMI).

The middleware server is the core of the system. It is responsible for extracting the metadata and storing it in the different layers which compose the metadata hierarchy. It consists of the metadata generator, which extracts metadata from input objects at the collection, ontology and object levels; the semantic manager, which consists of semantics extractors and similarity comparison modules to enable content-based retrieval. This server is written in Java and communicates with the object relational database system (ORDBMS) via the Informix JAI (Java Application Interface) (Informix, 1997).

The object relational server is responsible for the management of both metadata and data. The built-in data types and extensibility of the underlying ORDBMS enable management of all spatial objects. The multimedia objects are represented as 'smart large objects' which enable retrieval of the whole object or part of it. Multidimensional indexing is used to provide better performance in accessing the data.

To evaluate some of the ideas discussed in the preceding sections a simple prototype was implemented and tested. The prototype consists of an application that provides web-based access to a database of Ordnance Survey data through the Internet. The data set consists of the geometry and associated attributes of selected features of the city of Canterbury, i.e. a collection about the city of Canterbury with metadata referring to buildings, roads, railways, boundaries, etc. The client is responsible for map display and interaction with the user.

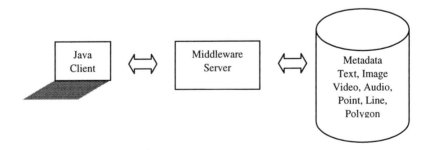

Figure 3.4 The prototype architecture

Figure 3.5 shows an example of the client GUI, with a feature selection capability including scrolling, display magnification or reduction, querying a particular region based on the Northings and Eastings, selection of the feature preferences, and palette preferences to assign colours to specific features. There is also a server log file which records all communications between clients and the middleware server. This log contains information about each client that connects to the server and the communications between the two. It can also be used to monitor requests. The prototype demonstrates that using a Java-based solution enables the system developer to determine the functionality provided to the user and to achieve an optimal distribution of execution load between the client and the server.

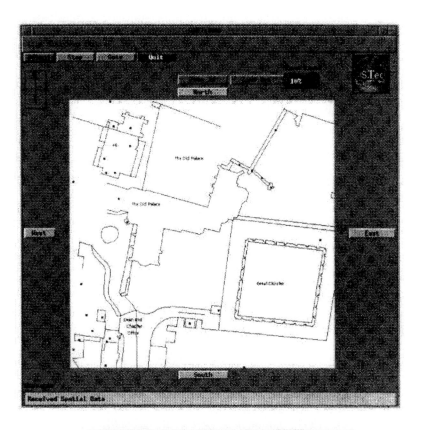

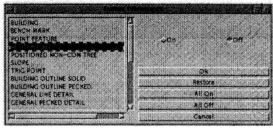

Figure 3.5 An example of the client GUI and the feature preferences dialog box

3.5 FUTURE DIRECTIONS

The prototype implementation is proof of the feasibility of the concept. What the current system also pinpoints are the problems awaiting solutions. Effective tailored application domain dependent systems can be provided for identified user communities. However, there is also a corresponding need for interoperability and standards not only in translating between different data models but also semantic concepts as well. Attention should be paid to the construction and use of ontologies at this level which consist of a common domain-derived vocabulary and possible relationships between the concepts captured therein. It enables users to express their queries at a high level of abstraction. The effective modelling of this ontological level is an interesting open research problem which is being investigated by geoscientist, artificial intelligence and database research communities.

The multiplicity of metadata standards require mapping and transformation capabilities to cope with the different metadata conventions. Providing centralised repositories of metamodels may enable different libraries to operate in a federated manner. In the geospatial domain the requirement for increasing spatial analysis and data visualisation functionality has major implications for performance. An interesting direction involves investigations into how high level knowledge-based concepts can be used to enable knowledge sharing, knowledge discovery and data mining over the Internet. The use of mediator or facilitator agents is one mechanism that would make it possible to share and fuse knowledge at a high level of abstraction.

ACKNOWLEDGEMENTS

The first author would like to thank the CAPES-Brazil for partially funding this research. The authors would like to thank Informix Inc. for the generous provision of a license to use the Informix Universal Server, under their University Grant Program.

REFERENCES

Andresen, D., Yang, T., Egecioglu, O., Ibarra, O. and Smith, T., 1996, Scalability Issues for High Performance Digital Libraries on the World Wide Web. In *Proceedings of ADL 1996, Forum on Research and Technology Advances in Digital Libraries*, IEEE, Washington DC, USA, pp. 139-148.

ANSI/NISO Z39.50, 1995, *Information Retrieval (Z39.50): Application Service Definition and Protocol Specification*, Z39.50 Maintenance Agency, Official Text for Z39.50.

Bergman, L, Castelli, V. and Li, C., 1997, Progressive Content-Based Retrieval from Satellite Image Archives. In *D-Lib Magazine*.

Best, C., 1997, Geospatial Profiles of Z39.50. In *Proceedings of the International Society for Photogrammetry and Remote Sensing Workshop: from Producer to User*, Boulder, Colorado, USA.

Buehler, K., 1996, What you need to do to prepare for an open GIS: working in an open environment, Geographic Information towards the Millenium. *In Proceedings of Association for Geographic Information AGI'96*, Birmingham, UK.

Burrough, P., Craglia, M., Masser, I. and Salge, F., 1996, Geographic Information: The European Dimension, *Discussion document from the European Science Foundation's GISDATA programme*.

Clement, G., Larouche, C., Gouin, D., Morin, P. and Kucera, H., 1997, OGDI: Toward Interoperability among Geospatial Databases. *ACM SIGMOD Record*, Vol. 26(3), pp. 18-23.

Ehmayer, G., Karppel, G. and Reich, S., 1997, Connecting Databases to the Web: a Taxonomy of Gateways. In *Proceedings of the Database and Expert Systems Applications, 8th International Conference, DEXA'97*, Hameurlain, A. and Tjoa, M. (eds) (Toulouse: Springer Verlag), pp. 1-15.

FGDC 1994, *Content Standards for Digital Geospatial Metadata Workbook*, Version 1.0, Federal Geographic Data Committee.

Franklin, M., Bjorn, J., Jónson, T. and Kossmann, D., 1996, Performance Tradeoffs for Client-Server Query Processing, *Proceedings of ACM SIGMOD Database Conference*, Montreal, Canada, pp. 149-160.

Güting, R., 1994, An Introduction to Spatial Database Systems. *Very Large Databases Journal,* Vol. 3(4), pp. 357-399.

Hamilton, G., Cattell, R. and Fisher, M., 1997, *JDBC Database Access with Java – A Tutorial and Annotated Reference*, (Reading: Addison-Wesley).

Informix, 1997, Java API Programmer's Guide, Version 1.04, (Menlo Park: Informix Press).

Kjeldsen, A., 1997, The CEO Enabling Service Search Subsystem. In *Proceedings of the International Society for Photogrammetry and Remote Sensing Workshop: From Producer to User*, Boulder, Colorado, USA.

Li, C., Bree, D., Moss, A. and Petch, J., 1996, Developing internet-based user interfaces for improving spatial data access and usability. In *Proceedings of the 3rd International Conference/Workshop on Integrating Geographic Information Systems and Environmental Modeling*, Santa Fe, New Mexico, USA.

Lockemann, P., Kolsch, U, Koschel, A., Kramer, R., Nikolai, R., Wallrath, M. and Walter, H., 1997, The Network as a Global Database: Challenges of Interoperability, Proactivity, Interactiveness, Legacy. *In Proceedings of the 23rd Very Large Databases Conference*, Athens, Greece, (Morgan Kaufmann), pp. 567-574.

Newton, A., Gittings, B. and Stuart, N., 1997, Designing a scientific database query server using the World Wide Web: the example of Tephrabase. In *Innovations in GIS 4*, Kemp, Z. (ed) (London: Taylor & Francis), pp. 251-266.

Melton, J. (ed), 1995, ISO/ANSI Working Draft, Database Language SQL3.

OGC, 1998, The OpenGIS Specification Model – Topic 11: Metadata, version 3.1, *OpenGIS Project Document Number 98-111r2*.

Ogle, V. and Stonebraker, M., 1995, Chabot: Retrieval from a Relational Database of Images, IEEE *Computer*, Vol. 28(9), pp. 40-48.

Smith, J. and Chang, S., 1996, VisualSEEK: a fully automated content-based image query system. In *Proceedings of the ACM Multimedia 96*, Boston, USA, pp. 87-98.

4

Web-based geospatial data conversion tools

Anup Pradhan and Bruce M. Gittings

4.1 THE INTERNET IS CHANGING GIS

One of the primary expenditures of implementing a Geographical Information System occurs in the data collection and base mapping stages (Korte, 1996). The data required by a particular individual or organisation could in fact already exist somewhere; the problem is trying to find it. With the advent of client/server communications, the Internet has become a powerful tool to transfer large amounts of information within the GIS user community. No other technology this decade has had such a profound impact on the computer and software industry of which GIS is a part.

This chapter will examine some of the issues responsible for the changes in the GIS industry due to the advent of web technologies. Primary emphasis is on a server-based system that merges existing technologies such as an HTTP server, a relational database and a GIS into a unified system that facilitates the online access and conversion of geospatial data over the Internet. This system is called the Geospatial Metadata Server (GMS) (Pradhan, 1996).

Each component of the system provides possible solutions to very real problems in accessing on-line digital data. At present, non-standard methods of cataloguing are usually used in providing descriptive information for on-line geospatial data. Lack of standardisation poses a serious problem for users searching for data based on some search criteria as well as for on-line data providers who wish to incorporate their collections with those that exist on other web servers. Efficient use of standardised data description methods is also an issue for on-line data providers, because the complexities of many current methods make cataloguing a very labour intensive task. Most search-engines used to access on-line data sets provide a bare minimum of the sort of functionality that can be achieved with present web technology. Finally, the Web has exposed users to geospatial data that can occur in a variety of different formats. The user, as a result, is either forced to acquire third party conversion software or purchase more than one GIS.

4.2 A STANDARDISED METADATA DATABASE

There are very few web-based examples of complete or even partial on-line implementations of the CSDGM standard (Federal Geographic Data Committee, 1994). Even with GMS only about 30 percent of the standard is implemented, however this was more due to time constraints imposed on the project rather than problems of complexity. On-line metadata usually exists as CSDGM-compliant text or HTML files, or is contained in an RDBMS (Relational DataBase Management System), using an ad-hoc schema providing only a few descriptive fields, which comprise the bare information minimum required to adequately describe a geospatial data set. Using ad-hoc database schemes to store metadata will inevitably preclude such sites from integrating their indexing and query capabilities with that of other sites, because each site will be using locally developed database schemes, that are incompatible with each other. The table and column names as well as their linkages would vary from site to site forcing the development of customised search engines.

The problem with sites that allow users to query entire metadata text or HTML files is that the files are excessively large providing far more information than is initially needed. Geospatial data sets are very complex pieces of software requiring a considerable amount of detail for a full description of their contents. That is precisely why detailed metadata standards such as CSDGM were developed. Yet users would have to sift through large quantities of metadata buried within individual files to find the required information. Specific sections and fields within individual metadata files would each need to be examined before a decision can be made as to the appropriateness of using any particular data set. The required information can be parsed from the text or HTML files, but this would require a considerable amount of time and processing. Therefore, at present the only effective method of indexing and querying large quantities of information is by using an RDBMS with a database designed on a specified standard (CSDGM) and query language (SQL). The interoperability of database schemes between different geospatial data dissemination sites and the ability to quickly query large quantities of metadata are the two primary reasons for using a RDBMS schema for GMS. This schema is called the GMS metadatabase.

4.3 UPLOADING, PARSING AND DATABASE ENTRY OF METADATA TEXT FILES

Metadata files are relatively large descriptive documents that are difficult to produce and just as difficult to store in a database. The FGDC metadata standard is designed to describe all possible geospatial data. There are 334 different elements in this standard of which 119 are compound elements that are composed of other compound elements in a hierarchical structure representing various aspects regarding the data set (Schweitzer, 1996). These elements are divided into seven major sections containing a number of minor sections, some of which are interconnected. Thus, the amount of information contained in any single metadata file is considerable. Spatial data providers invest a considerable amount of time and effort in the creation of metadata in order to provide the information to effectively use their data. The ability to parse metadata files and load the information directly into a database would be of considerable benefit to spatial data providers. The GMS metadatabase incorporates two utilities, MPARSER and MLOADER for this very purpose.

MPARSER initially scans a metadata file character by character and parses out the relevant sections, elements and element values and stores the results in a text file. The resulting text file is read into the MLOADER database entry program. MLOADER is not one but many smaller programs designed to enter metadata into each of the respective tables and columns of the metadatabase. All of the programs were written in ProC*, which is ORACLE Corporation's pre-compiler which allows one to embed SQL (Structured Query Language) code into C programs. MLOADER loads the information into their respective database tables, column by column and row by row maintaining referential integrity. Figure 4.1 describes the entire process.

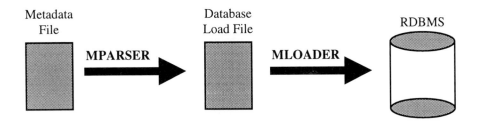

Figure 4.1 MPARSER and MLOADER within GMS.

Storing metadata within an RDBMS substantially increases the flexibility of search and indexing operations, that could not otherwise be achieved using less sophisticated non-proprietary database solutions. An RDBMS provides powerful facilities for the storage, query and report of metadata. Because organisations devote a considerable amount of time and effort in producing geospatial data, standardised metadata provides the means to advertise their holdings to an increasingly wider user community. Many Internet based data dissemination systems rely on the predictability of form and content that the CSDGM standard provides. The standard ensures that data providers and data users both speak the same language and arrive at the same understanding when accessing geospatial data on-line (Schweitzer, 1996). The use of an RDBMS gives the on-line data provider the freedom to decide:

- How the data should be searched?
- Which metadata sections or elements?
- How much information should be provided to the user?
- How should this information be structured or formatted?
- How should it be viewed?

Though the Web now has a number of on-line catalogues referencing geospatial data, none have devised a system whereby large amounts of metadata can be entered into a relational database via algorithmic parsing which can be searched on multiple criterions.

The National Geospatial Data Clearinghouse uses the Z39.50 protocol, that allows clients to simultaneously pass a single query to multiple servers (Nebert, 1995). The results from each server are compiled into a single HTML page, with links to on-line metadata, browse graphics and geospatial data. The documents returned are static and their metadata elements are not part of any relational database. In this case, querying to the servers can only be temporal, location or keyword based. Some of the NSDI nodes have Java applets that allow panning and zooming through layered map images but essentially it a modified version of a location based search engine (FGDC, 1996). Whereas, by implementing the CSDGM metadata standard in a relational database, queries can be conducted on any table, column or row so long as the queries are supported by the user interface. Additional types of queries can easily be added subject to user feedback. This is an important point because it is now prevalent that an increasing number of data providers are using the Internet as a medium to distribute their geospatial data. As the shear number of data sets increases on the Web, standard query mechanisms described above will no longer be useful on a national or international basis. Such sophisticated queries are only possible if the relational database on which the queries are being conducted is appropriately modelled and sufficiently expressive of the metadata standard.

4.4   ON-LINE METADATA ENTRY, RETRIEVAL, UPDATE AND REMOVAL OF DATABASE ROWS

GMS provides facilities for users to add their own geospatial metadata to metadata that already exists on the system. Such facilities have been a common feature of other web sites. The metadata is entered into an HTML form or Java programmed interface and subsequently indexed. Cataloguing in this fashion ensures the metadata can be queried on-line by a variety of mechanisms. Some sites like NASA's Global Change Master Directory (NASA, 1996) allows the user to create their own entry forms that submit metadata attributes compliant with the standard directory interchange format (DIF). Other sites have web-based metadata entry systems that provide a single HTML form to submit metadata compliant with the CSDGM standard! GMS, on the other hand, goes further by uploading metadata from the user's web site and inserting the information into its database. URL addresses referencing metadata files are either supplied by the user or collected using a web robot.

The amount of information contained within a metadata file that is necessary to appropriately represent one or more data sets is well beyond what the producer of the digital data is willing to enter in a sitting. What is needed is a method by which users can submit formally structured and compliant metadata to on-line geospatial data cataloguing sites in a way that divides the arduous task of forms based textual entry into several separate parts. One method would be to distribute a freeware/payware software application that allows users to write metadata into fields provided by the application. The results are saved in a proprietary document format which can then be continually opened, edited and saved, much like a word-processing document, until the entire metadata file has been entered.

The main problem with this approach is the software must be distributed to the user community at large, it must be supported on a variety of different computer platforms and it must be continually distributed in the event of periodic updates. This is a lot to ask considering the users traditional reluctance to add new software to their systems' because of problems associated with software installation, maintenance and eventual removal. Imagine doing this for every subscribed on-line data service.

Web-based solutions to metadata entry circumvent some of the problems just described with stand-alone application programs. The main issue in such implementations is the need for storage or persistence while the metadata is entered, updated or deleted from the database. GMS provides a series of programs that allows users to select, insert, update and delete metadata from the metadatabase on-line. A single metadata file is designated by a row number of the main metadatabase table, in this case, *Identification_Information* (II). All other RDBMS tables are linked by one-to-one, one-to-many and many-to-many relationships. User access to any particular row is controlled by a username and password entry system whose values are stored in the II table just mentioned and retrieved for periodic verification.

In all, the database is composed of 80 field elements or columns spread over 14 relational database tables and seven look-up tables. Each table is associated with its own HTML forms interface, CGI script and C processing program. Access to any of the 14 tables is through the same username and password system described above (see Figure 4.2).

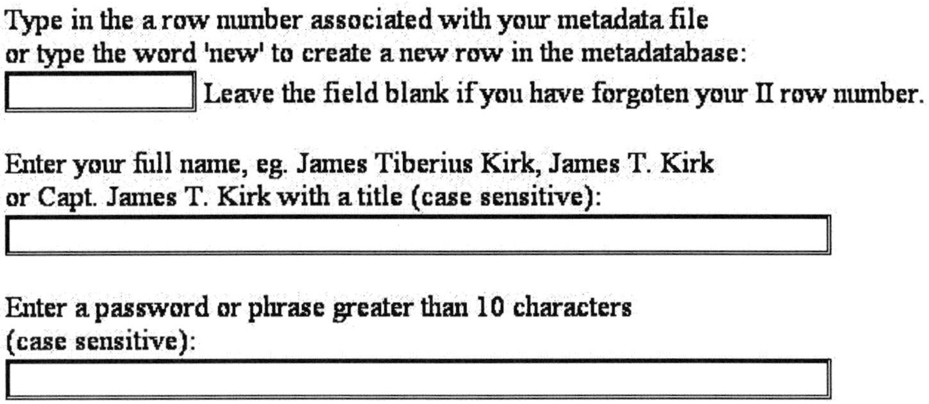

Type in the a row number associated with your metadata file
or type the word 'new' to create a new row in the metadatabase:

|_____| Leave the field blank if you have forgoten your II row number.

Enter your full name, eg. James Tiberius Kirk, James T. Kirk
or Capt. James T. Kirk with a title (case sensitive):

|_____|

Enter a password or phrase greater than 10 characters
(case sensitive):

|_____|

Figure 4.2   Row number and identification fields for the online metadata entry utility

Database rows can be added or subtracted, viewed and/or changed on any fourteen of the metadatabase tables depending on its relationship to the II table. Each metadata file is assigned a single row (Figure 4.2) in the metadatabase starting with the II table. To begin entering a metadata file, a new row is created in the *Identification_Information* table by typing the word 'new' into the above form when asked for row number. Alternatively, all the existing rows associated with a particular username and password can be displayed by simply leaving the 'row number' field blank.

4.5   THE CONVERSION OF GEOSPATIAL DATA REFERENCED BY METADATA

The geospatial data formats used in most GIS packages today are proprietary in nature, meaning the formats are designed by a specific GIS vendor and implemented for use within their own family of software products. These formats are not compatible with GIS packages from other vendors unless they are converted to a data format specified for the system in question. Incompatible geospatial data formats are probably the most prevalent cause of interoperability between the systems' of different GIS vendors. There are numerous data converters or translators to bridge the interoperability gap. Most vendors provide such conversion programs with their software but by no means do they cover all eventualities by incorporating all the formats that currently exist in the industry. Also, new data formats are continually being developed forcing the users of present GIS software into the very expensive option of upgrades. What about those users who do not have access to off-the-shelf GIS packages and require digital data for use in some third party software? Purchasing an entire GIS software package for just a few uses seems a rather expensive option. There are freeware data conversion utilities that can be downloaded from the Internet but frequently these utilities have limited documentation, are difficult to install, are difficult to use and may not run on the required platform.

One solution is to offer server side data conversion utilities over the Internet that can be easily accessed through any standard web browser. The system constructed for GMS uploads the geospatial data from any Internet site, converts it to a specified format and returns it to the user for download via a hypertext link. Problems of interoperability disappear because the Internet is being used as middleware to bridge the differences between client and server side systems. The user need not purchase expensive software nor install equally expensive and time consuming upgrades. Currently, GMS has three data conversion utilities which include, conversions to and from Arc/Info's interchange file format (E00) and AutoCAD's drawing interchange file (DXF), USGS's Digital Line Graph (DLG) and Topologically Integrated Geographic Encoding and Referencing System (TIGER), respectively. Conveniently and pragmatically, the data conversion utilities are implemented using Arc/Info's data conversion commands.

In all, there are six conversion utilities each comprising of several components. The web-based utilities were named after the data conversion commands they implement, which are outlined with their arguments in ESRI's help documentation shown in Figure 4.3.

What is unique about these geospatial data conversion utilities is that traditional data conversion technologies as seen with the Arc/Info commands (Figure 4.3) have been merged with current Internet technologies. Since data files must be accessed beyond the confines of the local operating system from another web server, the only conceivable reference these files can have in this context are URL (Universal Resource Locator) addresses. The use of URL addresses is in contrast to the original function of the underlying Arc/Info commands that access only local data files.

The development of an algorithm that parses URL addresses extracting the file names was necessary in order for the Arc/Info conversion commands to operate over the Web. Before uploading the data set, GMS checks if the resource is located within its data cache. The data cache is nothing more than a reserved directory that contains the most common geospatial data sets. The file names within the cache are compared to the file name extracted from a particular URL address. If the file already exists on the system, it need not be retrieved from the network thereby saving time and bandwidth. If the file must be retrieved from the Internet, then standard HTTP and FTP clients are used. The HTTP client consists of server side software called FETCH_HTTP by Jones (1995) who also wrote the DECthreads HTTP server on which GMS operates. Converting E00 to DXF, DLG or TIGER/Line file formats (ARCDXF, ARCDLG, and ARCTIGER) requires entering URL addresses of anywhere from one, two or three different coverages. So there is a possibility that these coverages can be uploaded from three different Internet sites. It is a clear demonstration of how GIS applications can access digital data beyond the confines of the local server from a seemingly limitless virtual network of databases.

E00 ←→ DXF
ARCDXF <out_dxf_file> {in_line_cover} {in_point_cover} {in_annotation_cover} {decimal_places}
      {ASCII | BINARY}
DXFARC <in_dxf_file> <out_cover> {text_width} {attrib_width}

E00 ←→ DLG
DLGARC <STANDARD|OPTIONAL> <in_dlg_file> <out_cover> {out_point_cover}
      {NOFIRST|ALL|ATTRIBUTED} {x_shift} {y_shift} {category}
ARCDLG <in_cover> <out_dlg_file> {in_point_cover} {in_projection_file} {x_shift} {y_shift}
      {in_header_file} {TRANS | NOTRANS}

E00 ←→ TIGER
TIGERARC <in_tiger_file_prefix> <out_cover> {out_point_cover} {out_landmark_cover}
ARCTIGER <out_tiger_file_prefix> <in_line_cover> {in_point_cover} {in_landmark_cover}

Figure 4.3 Arc/Info data conversion commands used in GMS

The conversion utilities just described form a system called a data-switch-yard whereby various GIS data formats can be converted to a neutral data format and converted once again to another desired data format (Figure 4.4). In this case the neutral data format is Arc/Info's interchange file format. However, other formats such as SDTS (Spatial Data Transfer Standard) can also be used. Perhaps more importantly, the utilities discusses in this section demonstrate that GIS functionality need not be restricted to the local database and operating system where the software resides.

With URL addresses to reference the location of a specific data file or files, it is possible to access and use geospatial data residing on any connected server. In addition to the local database, the network provides an added dimension for GIS applications to access and use geospatial data.

4.6 INTEGRATION OF METADATA DATABASE AND CONVERSION UTILITIES

The on-line data conversion utilities described in the previous section were initially developed as stand-alone utilities. By developing a database search engine, these utilities were integrated into the GMS metadatabase by querying database records and feeding the information retrieved directly into the conversion utilities. This section explains how

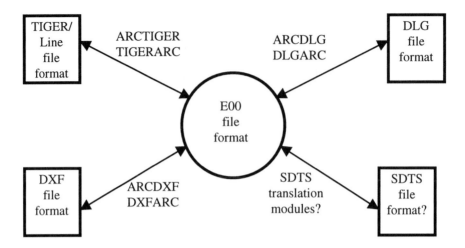

Figure 4.4 The data-switch-yard architecture

such on-line GIS software can be tailored to respond automatically to input from a web search engine.

The ability to query metadata files from a web site on the basis of date, keyword and/or co-ordinate information using HTML forms or Java applets is long established. The NSDI nodes for the FGDC Entry Point to the Geospatial Data Clearinghouse operate on this principle by providing an interface to a network of inter-connected data servers able to gather and distribute geospatial data and metadata over the Web. However, the limited search capabilities currently available may not be sufficient for databases (distributed or otherwise) that contains thousands or tens òf thousand of metadata files. Effective search engines must be able to query on a broad spectrum of metadata elements such as 'cloud cover' in satellite imagery or even 'fees', 'publisher', 'source' and/or 'citation'. This is precisely where an on-line metadata catalogue based on one or more metadata standards and implemented in an RDBMS comes into play. The GMS search engine can provide the powerful search capabilities just mentioned using any of its metadata elements to perform database queries on multiple criterions. The queries can be sent to multiple data servers using the same database scheme as GMS. The results can be collected/formatted and then shown to the client.

What is needed to integrate the conversion utilities and the metadatabase is certain information regarding digital data transfer. Since the information must come from the metadatabase in response to an on-line database query, the GMS search engine is capable of accessing this information because the database is designed in part on the CSDGM standard that has a section specifically describing and referencing on-line digital data. The table shown in Figure 4.5 constitutes a portion of the entity-relationship model of the GMS metadatabase, which corresponds to a section called the *Digital_Transfer_Information*. The columns of particular interest to the GMS search engine and conversion utilities are *Format_Name and Network_Address*.

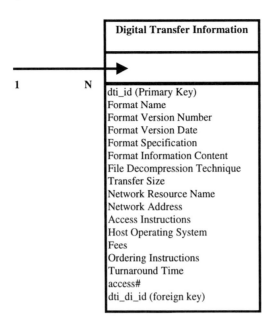

Figure 4.5 The Digital Transfer Information table

The information from the *Digital_Transfer_Information* table is accessed each time a metadatabase record is queried. GMS is capable of querying metadatabase records by one or more of the following mechanisms:

- a beginning and ending date;
- keyword or phrase;
- co-ordinates - circular search radius;
- co-ordinates - rectangular bounding area;
- co-ordinates - quadrilateral bounding area.

Extending the HTML interface to incorporate additional query mechanisms is a relatively simple task. Because the search engine is developed using the C programming language with imbedded SQL statements, accessing the metadatabase is considerably faster than using CGI scripts and the general purpose SQLPLUS utility. As a result, the performance of database queries is maximised. For database queries over a bandwidth-starved Web, this performance advantage is crucial. ORACLE's ProC* development environment allows the developer to construct database search engines that is not otherwise possible using standard SQL statements.

Metadatabase queries on date, keyword or co-ordinate criteria can be conducted, either individually or simultaneously, by the return of a UNION of database records. UNION is an SQL set operator that returns all distinct rows selected by either query (ORACLE7 Server, 1992). The location related queries are processed through C subroutines and then related back to the date, keyword and co-ordinate SQL queries. The user has the option of bounding a geographic area with a circle, a square/rectangle or a quadrilateral. The user can define a circular search area by submitting the latitude and longitude co-ordinates of the circle's central point and radius. Four bounding co-ordinates that represent the east, west longitudes and north, south latitudes delineate a search area that defines a box bounding area. Eight sets of co-ordinates or four separate lines that are not right angles to each other define a quadrilateral bounding area. The CSDGM standard defines the boundaries of geospatial data sets by an east and west longitude and a north and south latitude; essentially it defines the bounding co-ordinates of a square or rectangle.

A location related query would need to determine if the search area intersects any of the bounding co-ordinates of a particular data set. The GMS search engine computes this intersection using three subroutines, one that mathematically determines, a circle/line intersection, a line/line intersection and the distance between two sets of co-ordinates. The bounding co-ordinates of metadata records form a square or rectangle with each side representing finite lines. In a circular search query each of these finite lines are analysed to see if they intersect with the defined circle. In a quadrilateral search query the four finite lines are analysed to see if they intersect any of the four lines that define the sides of the quadrilateral. If intersections are found, then the primary key of the *Identification_Information* table, that uniquely defines a particular metadatabase record, is stored in an array for use later on in the program.

The box search query is constructed entirely of SQL statements along with keyword and date queries that follow after the UNION statement shown in Figure 4.6.

```
select IDENTIFICATION_INFORMATION.II_ID
    from ops$anp.identification_information
    where to_number(TL_BOUNDING_COORDINATE) <= :east
    and   to_number(TR_BOUNDING_COORDINATE) >= :west
    and   to_number(BR_BOUNDING_COORDINATE) <= :north
    and   to_number(BL_BOUNDING_COORDINATE) >= :south

    UNION
```

Figure 4.6   SQL statements for the box search query

There is a database query for each type of keyword (temporal, theme, place and stratum) whose specifications are outlined in the CSDGM standard. Each of the keyword queries and the date query is tied together with UNION statements to the above SQL code. Initially, only the primary key of the Identification_Information table (labelled as II_ID) is retrieved however this query sets off a series of loops that retrieves information of a particular metadata record from the Citation_Information, Browse_Graphic and Digital_Transfer_Information tables.

In addition to providing information on the transfer of on-line geospatial data, a number of metadata files contain hyperlinks to online graphic files that illustrate the types of products that can be constructed if one were to use the referenced data. This information is contained within the Browse_Graphic table and includes the file name, description, file type and online linkage (Figure 4.7).

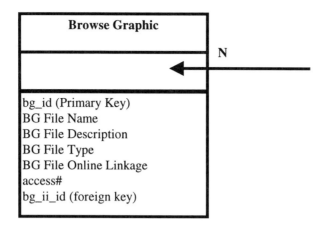

Figure 4.7   The Browse_Graphic table from the GMS metadatabase

The search engine is able to detect if this information exists within the Browse_Graphic table, which then creates HTML button(s) that are used to browse on-line graphic files.

```
printf("\n<form ACTION=\"http://www.geo.ed.ac.uk/anpexe/query.com\" METHOD=POST>");
printf("<input TYPE=\"submit\" VALUE=\"Browse Graphic File\">");
printf("<input TYPE=\"hidden\" NAME=\"preview\" VALUE=\"nil\">");
printf("<input TYPE=\"hidden\" NAME=\"browse\" VALUE=\"browse\">");
printf("<input TYPE=\"hidden\" NAME=\"bgfilenm\" VALUE=\"%s\">", vcbgfilenm.arr);
printf("<input TYPE=\"hidden\" NAME=\"bgfiledes\" VALUE=\"%s\">", vcbgfiledes.arr);
printf("<input TYPE=\"hidden\" NAME=\"bgfiletyp\" VALUE=\"%s\">", vcbgfiletyp.arr);
printf("<input TYPE=\"hidden\" NAME=\"bgfileonllink\" VALUE=\"%s\">", vcbgfileonllink.arr);
printf("</form>");
```

Figure 4.8   An HTML executable button to brown on-line graphic files

All of the information from the metadatabase table is contained within hidden HTML fields of the executable button and is sent to the CGI script on pressing the HTML button (Figure 4.8). The script then produces an HTML page on the fly with an imbedded graphic file.

The same procedure is used for the data conversion utilities. After completing all the queries, extracting all the information from the relevant database tables and reconciling any conflicting metadata records from the location, keyword or date queries, the search engine uses the queried information to provide linkages to executable programs on the server. The *Format_Name* column of the *Digital_Transfer_Information* table is used to determine those data conversion utilities that are appropriate to use with the on-line geospatial data (referenced in the metadatabase as URL addresses). Figure 4.9 shows how the search engine accomplishes this.

```
Directing system output to the Web browser:

if (!strcasen(format, "DXF", 3))
{
  printf("\n<form ACTION=\".....\" METHOD=POST>");
  .

  .
  printf("</form>");
}
if (!strcasen(format, "DLG", 3)) { HTML executable button }
if (!strcasen(format, "TGRLN", 5)) { HTML executable button }
if (!strcasen(format, "ARCE", 4)) { HTML executable button }
if (!strcasen(format, "ROBOT", 5)) { HTML executable button }
```

Figure 4.9 Sample code from the metadatabase search engine

In compliance with the CSDGM standard, DXF, DLG and TGRLN in the column *Format_Name* are standard acronyms indicating the URL in the column *Network_Address* references data in those proprietary data formats (FGDC, 1994). On-line software can be accessed through HTML executable buttons to convert any of these formats to Arc/Info's E00 export format. The E00 export file format is represented by the ARCE designation.

Within the same statement block (Figure 4.9) the search engine provides executable buttons that decompresses E00 files (import e00 → export e00 without compression) and an on-line Arc/Info coverage plotting utility. The Arcplot utility simply decompresses an E00 file to a coverage, plots it to a desired format (CGM, EPS, ADOBE or GIF) and then allows the plot to be either viewed on the browser or downloaded as a file. The user is able to set the plot parameters to decide which features are plotted (arcs, labels, links, nodes, polygons, points or tics).

4.7 WEB AGENTS IN GIS

What would happen if the *Network_Address* column (Figure 4.5) does not reference a specific geospatial data set but in fact provides a URL address to a site that contains tens, hundreds or even thousand of data sets? The GMS search engine (through the last IF-THEN statement in Figure 4.9) provides access to a web robot just for this eventuality. The robot, called Geobot, will scan a site domain HTML page by HTML page for links to geospatial data looking at their file extensions. As each page is uploaded, the robot parses it for hypertext links, which match a particular site domain. Hypertext links that reference HTML pages are placed in a 'to do' list which the robot uses to traverse the site. Those hypertext links which reference geospatial data sets as well as certain graphic file formats are placed in a 'links' file that can be later viewed by the user. Geospatial data links are also entered into a database that can be queried much like other web robot sites. So at the end of a run the user would have a text file of URL addresses indicating where the robot has been and an HTML page of geospatial data links indicating what the robot has picked up along the way.

4.8 CONCLUDING REMARKS

This chapter has illustrated how a networked geospatial metadata server, that gathers its own information as spatial data sets, can be implemented. Standard metadata facilitates, cataloguing and data access over the Web are provided for both geospatial data providers and users. GMS provides an on-line parsing and database entry program that uploads CSDGM compliant metadata and places it directly into the RDBMS or metadatabase. There is also a facility to create new metadata files or modify previously uploaded ones, through a series of HTML forms, that mirror some of the sections of the CSDGM metadata standard. The use of an RDBMS for the construction of the metadatabase provides the GMS search engine with powerful search capabilities, using a wide range of search criterion. The search engine not only retrieves linkages to on-line data sets but is also capable of executing other web-based software utilities. In theory, the possibilities are enormous, because the geospatial data and the on-line software can exist on any web server.

For demonstration purposes, all of the software accessed by the search engine comprises the separate GMS components, which includes the conversion utilities, the plotting utility and the web robot. The three conversion utilities, which form a data-switch-yard, demonstrate the immense capability of using the Web to convert between different GIS data formats. Long established problems of interoperability between different GIS products are resolved because the Internet provides all the necessary middleware that extricates the user from having to purchase more than one GIS or installing complicated third party conversion software. For web sites that possess a massive quantity of metadata and geospatial data files, GEOBOT (also accessed via the search engine) provides an effective means to traverse a server of relevant hyperlinks based on user defined preferences.

Efforts to standardise geospatial metadata will increase in importance as national standards give way to international standards and data providers are forced to choose between national or international data dissemination strategies. The use of an RDBMS to store metadata records as opposed to flat files is an important consideration in constructing search engines. The effectiveness of a well-designed and expressive search engine is perhaps the most important aspect of such web sites. The ease by which users are able to search metadata records and access geospatial data libraries will determine the usefulness of future web-based GIS applications. Concurrent database queries to multiple servers are likely to be a feature of on-line data dissemination, processing and viewing software. Such developments may address the performance issues associated with web-based data dissemination software but by no means can they be completely eliminated. The sort of file sizes and processing times required by today's GIS operations will ensure that problems associated with the lack of network resources remain an important consideration for data providers. However, despite the implementation issues associated with systems like GMS, all the programming tools and freeware already exist to build very useful GIS web applications that can be comparable to what is found in industry. The technologies used in this research are now at least five years old. With the advent of a whole new set of web technologies (Mace *et al.*, 1998) on the computing horizon, it is conceivable in the future that an increasingly larger share of GIS analysis will occur over the Internet.

REFERENCES

FGDC, 1996, Entry Point to Geospatial Data Clearinghouse, http://130.11.52. 178/
   gateways.html .
Federal Geographic Data Committee. 1994, *Content standards for digital spatial
   metadata*, Washington DC: Federal Geographic Data Committee.
Graham, I.M.. 1996, *HTML Source Book, 2ⁿᵈ Edition,* (New York: John Wiley & Sons).
Jones, D. 1995, *Fetch_HTTP: A Program that connects to a HTTP server and perform a
   fetch of the URL,* jonesd@er6.eng.ohio-state.edu.
Korte, G., 1996, Weighing GIS benefits with financial analysis. *GIS World,* 9(7), pp.
   48–52.
Mace, S., Flohr, U., Dobson, R. and Graham, T., 1998, Weaving a better Web. *Byte
   Magazine,* Vol. 23(3), pp. 58–68.
NASA, 1996, The Global Change Master Directory, http://gcmd.gsfc.nasa. gov.
Nebert, D., 1995, Status of the National Geospatial Data Clearinghouse on the Internet.
   U.S. Geological Survey, http://h2o.er.usgs.gov/public/esri/pl96.html.
ORACLE7 Server. 1992, *SQL Language Reference Manual,* Redwood City, CA: Oracle
   Corporation.
Pradhan, A, 1996, Geospatial Metadata Server**,** http://www.geo.ed.ac.uk/~anp/gms/
   main.htm.
Schweitzer, P. 1996, *FGDC Metadata FAQ*, http://geochange.er.usgs.gov/pub/tools/
   metadata/tools/doc/faq.html.

5

Implementing an object-oriented approach to data quality

Matt Duckham and Jane Drummond

5.1 INTRODUCTION

The influence of the United States' National Committee for Digital Cartographic Data Standards (NCDCDS) upon research into data quality cannot be understated. The 1988 'Draft proposed standard for digital cartographic data' (NCDCDS, 1988) set out five elements of data quality: lineage, positional accuracy, attribute accuracy, logical consistency and completeness. These quality elements are now widely used and have dominated the horizons of data quality for geoinformation researchers and professionals across the world (Morrison, 1995).

The guiding principle in the development of error handling in geographic information (GI) science has been that of *fitness for use* (Chrisman, 1983). Fitness for use aims to supply enough information upon the error characteristics of a data set to allow a data user to come to a reasoned decision about the data's applicability to a given problem in a given situation. The concept of fitness for use is congruent with the NCDCDS concept of *truth in labelling,* which rejects the formulation of prescriptive and arbitrary thresholds of quality in favour of the provision of detailed information, in the form of the five quality elements. However, the widespread acceptance of the NCDCDS quality elements does not necessarily bear any relation to their suitability for this task: to describe the quality of any data set so as to allow a user to assess fitness for use. Little or no empirical work exists to determine how well the NCDCDS 'famous five' perform in different scenarios and so their applicability to all situations is, at the very least, open to question. It is argued here that the success of the NCDCDS proposal as a standard is not indicative of its usefulness as a basis for the implementation of what has been termed an *error-sensitive* GIS (Unwin, 1995). The aim of this chapter is to explore the implementation of an error-sensitive GIS using a new approach to data quality. The approach used is inclusive of existing data quality standards, whilst allowing database designers, rather than system designers or standards organisations, to decide how best to describe the quality of their data.

5.1.1 Exhaustiveness

Anecdotal evidence, at least, suggests the five NCDCDS elements of spatial data quality are not exhaustive. The International Cartographic Association (ICA) Commission on Spatial Data Quality accepts the NCDCDS quality elements, yet feels the need to augment these with semantic and temporal accuracy. In 'Elements of spatial data quality', Morrison (1995) points out that it was 'more important to place solid definitions of the seven components in the literature than to attempt to be totally complete at this time'. Whilst often accepting the NCDCDS core of quality elements, many researchers have suggested the use of a number of other data quality elements, such as source and usage (Aalders, 1996), detail (Goodchild and Proctor, 1997) and textual fidelity (CEN/TC287, 1996). Even a cursory examination, then, reveals considerably greater than five quality elements. Further, it is worth noting that there is no firm agreement on the actual definitions of many of the elements mentioned (Drummond, 1996). It is unreasonable to expect any data quality standard to be exhaustive; there will always be eventualities and situations for which the standard was not intended. Therefore, the design of error-sensitive GIS should allow enough flexibility to accommodate not simply existing data quality standards, but the redefinition of these standards and the use of novel, user defined data quality elements.

5.1.2 Meta-quality

The NCDCDS model of quality restricts quality information to referring only to geographic information: the concept of quality as *meta-data*. The term *meta-quality* is used to denote quality information about quality (Aalders, 1996). The European Committee on Standardisation (CEN) draft paper on geographic data quality contains a detailed description of one possible approach to meta-quality, defining four meta-quality elements. The draft proposes that confidence, reliability and methodology should be reported for quality information in addition to an abstraction modifier, which describes the degree of distortion of quality resulting from the abstraction of the real world to the ideal data set (CEN/TC287, 1996). Whilst the CEN proposed standard is more extensive than the NCDCDS standard, it still restricts itself to a number of predefined meta-quality elements and to a single level of self-reference. Once the concept of meta-quality is acknowledged as important, however, there is no particular reason to impose these restrictions. Preferable to the arbitrary prescription of meta-quality elements and structure would be an error-sensitive GIS design free from such limitations.

5.1.3 An inclusive approach to data quality

The intention here is not to downplay the importance of the five NCDCDS quality elements nor of quality standards in general. The extensive use of the NCDCDS standard in particular, in one form or another within national and international standards organisations, is an indication of its value. However, for the reasons of exhaustiveness and support for meta-quality, it is clear that such quality standards should not form the basis of a GI science approach to data quality. To illustrate, the development of GIS would have been very different had spatial database design restricted GIS users to a standard set of geographic objects. Similarly, restricting error-sensitive GIS users to only those quality elements based on data quality standards will inhibit the uptake, growth and efficacy of data quality for those users. Though scarce, previous attempts to implement error management within GIS have tended to be based on the NCDCDS quality elements (e.g. Guptill, 1989; Ramlal and Drummond, 1992; van der Wel *et al.*, 1994). Such an approach places the responsibility for deciding what data quality elements best describe the fitness for use of a data set on the standards organisations. The goal of an inclusive approach to data quality is to devolve this responsibility to the database designer, whilst still respecting the vital rôle of standards in reporting data quality.

5.2 BACKGROUND

This chapter draws on two existing strands of research, object-orientation (OO) and a conceptual model of information systems (IS) design, to formulate and implement an inclusive approach to data quality.

5.2.1 Conceptual model of information systems design

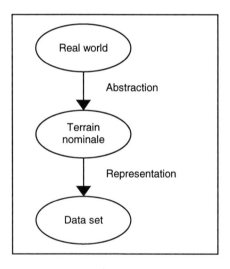

Figure 5.1 Conceptual model of IS design

Conventionally, the process of IS design has been conceptualised as summarised in Figure 5.1 (see David *et al.* (1996), Kainz (1995), Maguire and Dangermond (1991), Worboys (1992)). The infinite complexities of the real world can only be usefully handled following the abstraction of the real world to an ideal data set, termed the *terrain nominale*. An actual data set or database constitutes a representation of this terrain nominale. However, in contrast to geographic objects, there is no meaningful concept of data quality in the real world. Data quality is not usefully modelled by the process of abstraction and representation, rather it is only as a consequence of the deficiencies of these processes that data quality emerges at all (David *et al.*, 1996).

5.2.2 Object-oriented analysis and design

Object-orientation is increasingly the default choice for a wide range of information systems design. The fundamental features of OO are classification, encapsulation and inheritance. Classification deals with complexity in the real world from the top down and focuses on the features of an object in the real world that distinguish it from other objects. Complementary to classification, encapsulation approaches complexity from the opposite direction and aims to hide the detailed mechanisms and features of an object. Finally, inheritance is a way of structuring classified, encapsulated objects in an hierarchy that allows blueprint features to be described only once. Inheriting sub-classes can then incrementally specialise this blueprint.

The process of object-oriented analysis (OOA) uses classification, encapsulation and inheritance to create OO representations of the real world in an implementation independent way. It is a technique which aids the resolution of the enormous complexity of the real world to the relative simplicity of an information system. A modified version of OOA has been demonstrated to significantly improve the process and results of GIS development (Kösters *et al.*, 1997). Object-oriented design (OOD) is a complementary process which aids the creation of a particular implementation architecture from the results of an OOA. Object-oriented design contrasts with OOA in that the design process is implementation dependent and aims to produce results that can be programmed directly into a particular OO environment. A more detailed exposition of OOA, OOD and of OO generally is given in the literature (e.g. Booch (1994), Coad and Yourdon (1991a; 1991b), Rumbaugh *et al.* (1991), Worboys (1995)). Object-oriented analysis and design can be used with the conceptual model of IS presented above to tackle the problems posed by the incorporation of data quality into GIS.

5.3 OBJECT-ORIENTED CONCEPTUAL MODEL OF DATA QUALITY

A fundamental distinction between data quality objects and geographic objects, then, is that data quality only comes into existence as a result of the process of abstracting and representing the real world. From this standpoint two new object-oriented classes of quality can be identified: abstractive quality and representative quality. The following sections present these two classes in detail and indicate how they are arguably inclusive not only of the NCDCS model of quality, but of any reasonable set of quality elements.

5.3.1 Representative quality

The process of representation inevitably entails the introduction of error. Representative quality is data quality that records error introduced through the representation of the terrain nominale. An example is the NCDCS quality element positional accuracy, which is defined as the difference between an observed position and the 'true' position (NCDCS, 1988). The 'true' position can be found in the ideal data set, or terrain nominale. Similarly, the NCDCDS data quality element lineage records the process history of data and is often seen as the starting point for any data quality standard. Both of these definitions are consistent with the concept of representative quality elements.

Two additional features of representative quality are required to model the entire range of expected representative quality behaviours. First, some representative data quality elements are only meaningful when the data to which they refer is of a particular type or metric. For example, the concept of positional accuracy is only sensible when discussing spatial data. The quality storage model needs to be able to restrict the scope of a given quality element to defined metrics.

Second, a few data quality elements can be thought of as compound, whilst most can not. For example, a data object may be annotated with a constantly updated list of lineage objects detailing the different operations and events through which the object has passed. The same object can, in contrast, only have one source object. The former is an example of a compound data quality element, whilst the latter can be regarded as a special, restricted case of a compound data quality element. The compound behaviour of representative data quality objects also needs to be reflected in the OOA.

Crucially, representative quality operates at an object level rather than a class level. The OOA of representation suggests that representative quality is expected to vary from object to object. There is no particular reason why one object should have the same lineage, say, as a second object, even if they are of the same class.

5.3.2 Abstractive quality

We can never hope to formulate a terrain nominale that completely describes the real world. However, for a particular abstraction of the real world, it may be desirable to ensure that certain properties of the real world persist in the terrain nominale and so into the data set. Abstractive quality is defined here as data quality which supports or informs the linkage between the real world and the terrain nominale, links which would have otherwise been lost in the process of abstraction. The CEN quality element abstraction modifier is a good example of the concept. An abstraction modifier is a textual record of the distortion resulting from the process of abstraction (CEN/TC287, 1996) and is clearly an abstractive quality element. The NCDCDS quality element logical consistency is another example, although it is not particularly useful in an OO environment. Logical consistency is intended as a check on the logical content and structure of data. For example, logical consistency in the form of topological consistency can be used in contour data to check that the height of an individual contour falls somewhere between its topological neighbours and that contour lines do not cross. This behaviour can be enforced with the abstractive quality element logical consistency. However, an OOA of contours would usually encode behaviour within the contour class itself, negating the use of logical consistency in this example. Through careful use of OOA, logical consistency can effectively be removed from the data quality discussion.

In contrast to representative quality, OOA suggests that abstractive quality operates exclusively at the class level. Abstraction produces simplified classes of objects from complex real world objects. Any deficiencies in the process of abstraction will be felt equally by all objects of a particular class. For instance, the degree of abstraction given by an abstraction modifier is expected to be homogeneous across all objects of a particular class.

5.3.3   Quality storage model

Object-oriented analysis allows a sharp line to be drawn between class based abstractive quality and object based representative quality. This and other details of the quality storage model are illustrated in the class diagram in Figure 5.2. A class diagram gives classes, inheritance and aggregation used in an OOA in addition to any important behaviours on the classes, termed *methods*. The notation used in Figure 5.2 is based on that of Rumbaugh *et al.* (1991), and bold type is used in the following text to highlight class and method names.

The class **representative element** is the super-class of all representative quality elements and is a collection of objects of class **representative attribute**, which in turn define the individual attributes of a data quality element. The positional accuracy quality element, for example, might be constructed by creating a new class **positional accuracy**, which inherits from **representative element**. This class would be an aggregate of a number of new classes inheriting from **representative attribute**, for instance, **x-RMSE** and **y-RMSE**. Additionally, the **metric scope** of the **positional accuracy** class would be set to restrict positional accuracy objects to referring to spatial objects only. The **is compound** behaviour would be set such that at most one **positional accuracy** object could refer to the same geographic object. It is worth emphasising here that the choice and structure of representative data quality elements used is entirely free as long as the quality classes follow the basic pattern of a named collection of quality attributes with a metric scope and a compound behaviour.

The class **geographic objects** represents the super-class of all geographic objects in the terrain nominale. The class **uncertainty** has two methods, **get representative quality** and **set representative quality**. Since the class **geographic objects** inherits from the class **uncertainty**, every geographic object in the database will have its own **get representative quality** and **set representative quality** methods with which to access its own data quality. All **geographic objects** also inherit from **abstraction**. The **abstraction** class contains any number of methods that outline the supported abstractive quality elements. In Figure 5.2 the ellipsis below the **abstractive quality** method emphasise that other abstractive quality elements can be used. Because abstractive quality elements are inherited behaviours, they will be identical for all geographic objects of a particular class. This does not necessarily mean that behaviours are identical for all classes of geographic objects. Each method would usually be redefined or overridden for new geographic object classes. The **abstraction** class simply guarantees that all **geographic objects** have the 'hooks' on which to hang abstractive quality elements.

The requirement for all geographic objects to inherit from **abstraction** and from **uncertainty** is the only restriction placed on the geographic objects which can be represented in the database. Inheritance allows the transmission of error-sensitive behaviour to all sub-classes of **geographic objects**. The implication is that any OO database could be supported, including existing databases.

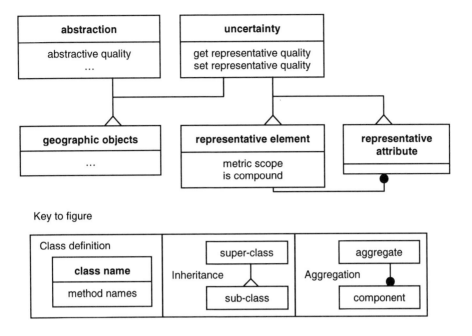

Figure 5.2 Class diagram of OOA results

5.3.3.1 Meta-quality

The classes **representative element** and **representative attribute** themselves inherit from **uncertainty**. This addition introduces meta-quality into the class structure; each **representative element** and **representative attribute** object will have **get representative quality** and **set representative quality** methods. Using this scheme it is possible to store quality information about quality. Further, this is not restricted to a single level of meta-quality. The system is recursive and so meta-quality can be assigned to arbitrary levels of self-reference. Such a system inevitably has built in redundancy and may be only infrequently of use; however, it is in keeping with the objective of placing as few prerequisites as possible upon what data quality models can be supported.

5.4 GOTHIC IMPLEMENTATION

The quality storage model resulting from the OOA described in the previous section was implemented in Laser-Scan Gothic object-oriented GIS. Gothic is one of the few commercial GIS which employs a fully OO database. The Gothic implementation was successful, first in that it was able to confirm the desired properties of the quality storage model. Second, while the Gothic database allows some flexibility it does employ its own data model, so the implementation provides some support for the implied compatibility of the quality storage model with existing OO databases. The technical details of coding and implementation are relatively straightforward and consequently not of interest to this discourse. However, some issues of more general interest are discussed in this section.

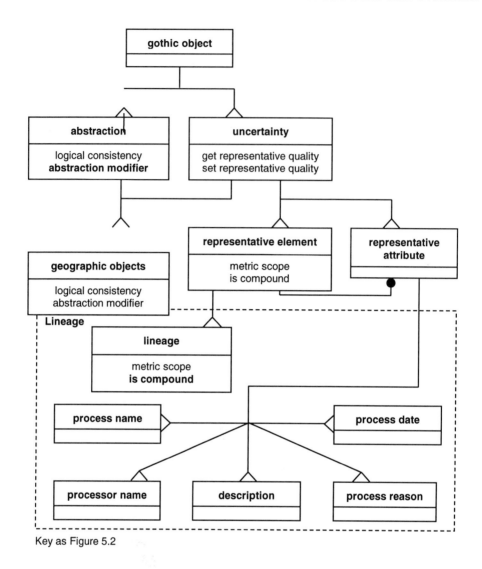

Key as Figure 5.2

Figure 5.3 Class diagram of Gothic implementation

5.4.1 Gothic quality storage model

Object-oriented design was used to implement the quality storage model in Gothic and produce the class diagram given in Figure 5.3. Very few changes to the OOA were needed in the OOD for Gothic; the core classes are identical to those results presented in Figure 5.2. To benefit from coding and support already within Gothic, every class inherits from a generic Gothic super-class, **gothic object**. The class **abstraction** has two abstractive quality elements, **logical consistency** and **abstraction modifier**. Figure 5.3

highlights that these two methods will be overridden by redefined **logical consistency** and **abstraction modifier** methods on any sub-classes.

The final addition to Figure 5.3 is the example lineage element. Aalders (1996) defines lineage is as a statement describing each process an object has undergone, including who undertook the process, when and why. This can be represented in the database by defining a **lineage** class, which inherits from **representative element**, and five sub-classes of **representative attribute**: **process name**, **processor name**, **process date**, **process reason** and **description**. The **metric scope** method would be overridden to reflect the unlimited metric scope of lineage objects and their ability to refer to any object in the database. Similarly, the overridden **is compound** would enable a number of different lineage objects to refer to the same target object.

5.4.2   Results of the implementation

By using this data quality storage model, database designers should be a position to represent data quality elements of their choice. Figure 5.3 gives an example of how this might be done with lineage. Clearly, any number of other data quality elements can be represented in this way. The details of what data quality elements are to be represented and what attributes each element maintains is also a matter for the database designer rather than the system designer or standards organisation. One possible set of data quality elements, derived from those suggested in the literature and incorporating the five NCDCDS elements of spatial data quality is given in Table 5.1. These quality elements were implemented in Gothic under the quality storage model presented here. The implementation was satisfactorily able to represent and store this range of quality elements for simulated geographic data.

5.5   DISCUSSION

The OOA analysis of data quality undertaken here revealed a novel approach to data quality storage within GIS that is not tied to any particular set of data quality elements. Following OOA, OOD was used to implement the OOA in Laser-Scan Gothic GIS, producing the core of an error-sensitive GIS. This error-sensitive GIS core was able to represent a suite of data quality elements taken from the literature and incorporating the NCDCDS five spatial data quality elements. The incorporation of existing data quality elements within the quality storage model is straightforward, and is taken to imply that any data quality element in the literature or that might reasonably be proposed is expected to be incorporated with similar ease. The movement of fundamental decisions about data quality away from standards organisations and systems designers and towards database designers is seen as central to combating the lack of support for data quality that is currently evident in commercial GIS. Since the design and implementation presented here does not preclude the use of any reasonable data quality element, it supports this shift in emphasis.

Table 5.1 Example data quality element implementations

Quality element	Quality attributes	Metric scope	Reference
Categorical accuracy	Percentage probability	Aspatial categorical data only	NCDCS
Continuous accuracy	RMSE	Aspatial quantitative data only	NCDCS
Completeness	Percentage present	Any	NCDCS
Currency	Last update Temporal validity	Any	CEN/TC287
Detail	Mathematical precision	Numerical data only	Goodchild and Proctor
Lineage	Process name Process date Process description Process reasons Processor name	Any	Aalders
Positional accuracy	x-RMSE y-RMSE	Spatial data only	NCDCS
Source	Source organisation Reason for creation Creation date	Any	Aalders
Textual fidelity	Percentage correct spelling	Textual data only	CEN/TC287
Usage	User name Usage type Usage date Usage comments	Any	CEN/TC287
Abstraction modifier	N/A	N/A	CEN/TC287
Logical consistency	N/A	N/A	NCDCS

5.5.1 Future work

Whilst this work does set out a core of data quality storage functionality, further steps remain before the implementation can truly be called error-sensitive. First, only the storage of data quality is discussed here. Error propagation and visualisation are also needed and are currently being implemented. Second, considerable advantages in storage volumes are possible using the quality storage model presented here and this is also now being implemented. Lastly, the model requires further testing with a much wider range of data quality elements and with real geographic data to verify its usefulness.

The error-sensitive GIS can appear quite complex and difficult to understand at times. This complexity is set to increase as propagation and visualisation modules are added. Further, the quality storage model places no restrictions upon whether the quality statistics used actually make any sense. For example, unless the database designer defines positional accuracy with a metric scope restricted to positional objects, there is nothing to stop the database user from assigning positional accuracy to, say, thematic attributes. This problem has wider significance: in addition to its potential advantages, the delegation of responsibility for defining data quality elements to the database designer carries the potential for more serious errors to be made. Finally, the storage of data quality in a database is only a means to an end: to assess the fitness for use of a given data set for a given use. Most data users would not expect to be able to interpret complex data quality information in a way that allows them to assess fitness for use. In order to address all of these issues, the long-term aim of this research is to produce not simply an error-sensitive GIS, but an *error-aware* GIS. The error-aware GIS currently under development makes use of knowledge-based technology and should be able to bridge the gap between error-sensitive GIS and the need for tools to allow non-specialist users to assess the fitness of their data for any particular use.

ACKNOWLEDGEMENTS

The Natural Environmental Research Council funds this research as a CASE studentship supported by Survey and Development Services (SDS), UK. Supervision for the research is provided by Dr David Forrest of Glasgow University Geography and Topographic Science Department and John McCreadie at SDS in addition to the second author on this chapter. Gothic software was kindly supplied under a development licence by Laser-Scan, UK.

REFERENCES

Aalders, H., 1996, Quality metrics for GIS. In *Advances in GIS Research 2: Proceedings of the Seventh International Symposium on Spatial Data Handling*, Vol. 1, Kraak, M. J. and Molenaar, M. (eds) pp. 501–10.

Booch, G., 1994, *Object-oriented analysis and design with applications*, 2nd., (Benjamin-Cummings: California).

CEN/TC287, 1996, Draft European standard: Geographic information–quality. Technical report prEN 287008, European Committee for Standardisation.

Chrisman, N., 1983, The role of quality information in the long term functioning of a geographic information system. In *Proceedings Auto Carto 6*, Vol. 1., pp. 303–312.

Coad, P. and Yourdon, E., 1991a, *Object-oriented analysis,* (Yourdon Press: New Jersey).

Coad, P. and Yourdon, E., 1991b, *Object-oriented design*, (Yourdon Press: New Jersey).

David, B., van den Herrewegen, M. and Salgé, F., 1996, Conceptual models for geometry and quality of geographic information. In *Geographic objects with indeterminate boundaries*, (Taylor and Francis: London), pp. 193–206.

Drummond, J., 1996, GIS: The quality factor. *Surveying World*, Vol. 4(6), pp. 26–27.

Goodchild, M. and Proctor, J., 1997, Scale in a digital geographic world. *Geographical and environmental modelling*, Vol. 1(1), pp. 5–23.

Guptill, S., 1989, Inclusion of accuracy data in a feature based object-oriented data model. In *Accuracy of spatial databases*, Goodchild, M. and Gopal, S. (eds) (Taylor and Francis: London), pp. 91–98.

Kainz, W., 1995, Logical consistency. In *Elements of spatial data quality*, Guptil, S. and Morrison, J. (eds) (Elsevier Science: Oxford), pp. 109–137.

Kösters, K., Pagel, B-U. and Six, H-W., 1997, GIS-application development with GEOOOA. *International Journal of Geographical Information Science*, Vol. 11(4), pp.307–335.

Maguire, D. and Dangermond, J., 1991, The functionality of GIS. In *Geographical information systems*, Vol. 1., Maguire, D., Goodchild, M., and Rhind, D. (eds) (Longman: Essex), pp 319–335.

Morrison, J., 1995, Spatial data quality. In *Elements of spatial data quality*, Guptil, S. and Morrison, J. (eds) (Elsevier Science: Oxford), pp. 1–12.

NCDCDS, 1988, The proposed standard for digital cartographic data. *American Cartographer*, Vol. 15(1), pp. 11–142.

Ramlal, B. and Drummond, J., 1992, A GIS uncertainty subsystem. In *Archives ISPRS Congress XVII*, Vol. 29 B3, pp 356–362.

Rumbaugh, J., Blaha, M., Premerlani, W., Eddy, F. and Lorensen, W., 1991, *Object-oriented modeling and design*, (Prentice Hall: New Jersey).

Unwin, D., 1995, Geographical information systems and the problem of error and uncertainty. *Progress in Human Geography*, Vol. 19(4), pp. 549–558.

van der Wel, F., Hootsmans, R. and Ormeling, F., 1994, Visualisation of data quality. In *Visualisation in modern cartography*, MacEachren, A. and Taylor, D. (eds) (Pergamon: Oxford), pp 313–331.

Worboys, M., 1992, A generic model for planar geographical objects. *International Journal of Geographical Information Systems*, Vol. 6(5), pp. 353–372.

Worboys, M., 1995, *GIS: A computing perspective*, (Taylor and Francis: London).

PART II

Integrating GIS with the Internet

6

The Web, maps and society

Menno-Jan Kraak

6.1 INTRODUCTION

Cartography is changing: changing from being supply-driven to demand-driven. More people will be involved in making maps. More maps will be created, but many of them only for a single use. These maps are changing from being final products, presenting spatial information, to interim products, that facilitate our visual thinking. Maps will be the primary tools in an interactive, real-time and dynamic environment, used to explore spatial databases that are hyperlinked together via the World Wide Web.

This process is being accelerated by the opportunities offered by hardware and software developments. These have changed the scientific and societal needs for geo-referenced data and, as such, for maps. New media, such as the Web, not only allow for dynamic presentation but also for user interaction. Users expect immediate and real-time access to the data.

This trend is strongly related to what has been called the 'democratisation of cartography' by Morrison (1997). He explains it as "...using electronic technology, no longer does the map user depend on what the cartographer decides to put on a map. Today the user is the cartographer". And "...users are now able to produce analyses and visualisations at will, to any accuracy standard that satisfies them....". Those of us active in geoinformatics might feel a little uneasy here. Are the data available and if so are they of sufficiently high quality? Are the tools to manipulate and visualise the data available at a 'fitness for the user' level? Without positive answers to these questions, map use may lead to the wrong decisions. However, it seems the trend is irreversible and will have a tremendous impact on our disciplines. A challenging question is 'what will remain for the cartographer and other geoinformatics experts?'

The changes require a different type of cartographer, someone who is an integral part of the process of spatial data handling. As well as traditional cartographic design skills, cartographers must make their knowledge available to fellow geoscientists using interactive real-time maps to solve their problems. For both the cartographer and the geoscientist the necessary data will be retrieved from all over the country, if not the world. The National Spatial Data Infrastructure will be the highway to travel to get to this data. Maps can also facilitate the use of this infrastructure by acting as the guide, en route, to the data. This chapter will address the following questions: What role can/will the Web play in cartography and geoinformatics and how can maps function on the Web to meet the user's needs?

The Web, among other technological developments, stimulates the emergence of a new kind of cartography: exploratory cartography. It will create an environment where geoscientists can work to solve their problems and make new discoveries. A challenge for cartographers is to create or improve mapping *tools* that allow exploration. They will put the map in its natural role as the *access* medium to the National Spatial Data Infrastructure. However, it remains the task of the cartographer to test the *effectiveness* of these maps (products). This chapter will examine what has been achieved with respect to calibration of the Web and exploratory cartography.

6.2 THE WEB AND MAPS: PRODUCER PERSPECTIVE

Traditionally, but still very much valid today, the cartographer's main task was to make sure good cartographic products were created. A major function of maps is still the transfer of *information* from spatial data; in other words to inform us about spatial patterns. Well-trained cartographers are designing and producing maps supported by a whole set of cartographic tools and theory as described in cartographic textbooks (Robinson *et al.*, 1995; Kraak and Ormeling, 1996).

During the preceding decades, many others, who are not cartographers, have become involved in making maps. The widespread use of Geographical Information Systems (GIS) has increased the number of maps created tremendously (Longley *et al.*, 1998). Even the spreadsheets used by most office workers today have mapping capabilities, although most of users are probably not aware of this (Whitener and Creath, 1997). Many of these maps are not produced as final products, but rather as intermediate resources to support the user in his or her work dealing with spatial data. The map, as such, has started to play a completely new role: it is not just a communication tool but also a tool to aid the user's (visual) thinking process. Increasingly, the Web is playing a more prominent role in this.

Initially, the Web was used simply as another medium to present spatial data. And for most maps found on the Web this is still the case. We can ask: 'Why are maps distributed via the Web?' The answer is relatively simple. Distribution is cheaper and faster, and allows for frequent updates. An additional advantage is that, while using the possibilities of the Web the maps can be made interactive. The interactivity and ease of use of the Web is currently very much limited by technology. However, developments are proceeding quickly, and what is not possible today certainly will be tomorrow. Increasingly those who offer maps and spatial data via the Web try to make them more accessible for exploratory activities on the user's side. The current options and limitations require to be reviewed.

In this context, the following questions are important: 'What is different about using the WWW as the media to display maps?' 'What can be done that could not be done before, and what can no longer be done?' 'How do we deal with these new options and the limitations the WWW offers?' 'What effect does it have on the user?' 'What are the needs from an exploratory cartographic perspective?'

6.2.1   Cartographic considerations

From a cartographic design perspective there are few differences between map design for the Web or any other form of on-screen maps. The map size (screen resolution) remains limited. However, an additional point regarding size has to be considered: This is the storage size of the map image, which is critical with regards to retrieval time. Users are not eager to wait very long, and if they have to, it is likely they will switch to another web site for their information. An alternative is that one offers the user different versions (in size and/or format) for different purposes. But this might reduce some of the benefits of map distribution via the Web for the providers, especially when there is a high degree of interactivity in the map. Because of these size constraints maps may need to have a limited information content only. However, the design must be attractive and informative, to compensate for waiting time.

The need for interactive tools in relation to exploratory cartography is significant (Kraak, 1998). Tools are required which allow the user to pan, zoom, scale, and transform the map. At any time, the user should be able to know where the view is located and what the symbols mean. They should have access to the spatial database to query the data. To stimulate visual thinking, an unorthodox approach to visualisation is recommended, which requires the availability of alternative visualisation techniques and links to multimedia components such as sound, text, video and animation. However, the Web options listed below do not offer this functionality yet. This is mainly due to technological barriers but also bandwidth.

6.2.2   Technical considerations

If one looks at the typical display options, it is possible to distinguish between several methods that differ in terms of necessary technical skills from both the user's and provider's perspective. The overview given below can only be a snapshot at the time of writing this chapter, since development on the WWW is tremendously fast. This 'classification' is certainly not carved in stone, and one might easily find examples that would fit these categories, or define new categories or combinations. Its objective is to give an overview of current possibilities, based on how the map image is used. The samples in the respective categories given in the appendix to this chapter are randomly chosen from the huge number of possibilities on the ever growing Web, and not necessarily the most representative (see Figure 6.1 and Appendix).

Bitmaps
Maps presented as static bitmaps are very common. Often the sources for these maps are original cartographic products, which are scanned and placed on the WWW. These images are view-only. Many organisations such as map libraries or tourist information providers make their maps available in this way and indeed, this form of presentation can be very useful, for instance, to make historical maps more widely accessible. The quality of the images depends very much on the scan-resolution.

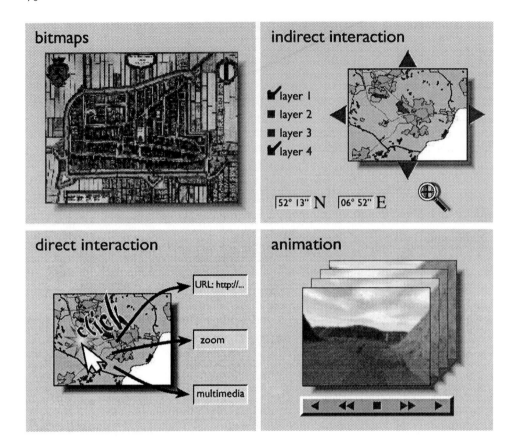

Figure 6.1 Technical options to display a map on the WWW, based on interaction with the map display

Indirect interaction with map image
Indirect interaction means that the user is not interacting with the map, but influences the map contents via a separate menu. Examples are menus to switch layers on or off, or dealing with map projections to determine the co-ordinates of the map centre. The final map result is still a bitmap, but the major difference with the previous category is that the user has influenced its contents. These 'maps-on-demand' can be final products or just a preview of what data can be downloaded or acquired by different means.

Direct interaction with the map image.
The simplest versions of maps in this category are the clickable maps. In this case a 'filter' is used, which defines an area in the image as hotlinked to a particular location. The map functions like a menu. Clicking an area will normally link the user to another URL with additional information. It is also possible to simulate pan and zoom by offering another map in response to the user's click.

More advanced direct interaction can results in GIS-like operations and allow one to access the data behind the map. The maps not only allow zooming and panning, but also querying. Some systems offer even more options and such maps, are often created via JAVA or VRML programming.

Animations
Animations are about change; change in one or more of the spatial data's components. On the Web several options to play animations are available. The so-called animated-gif is perhaps the simplest of the animations. It consists of a set of bitmaps, each representing a frame from an animation are positioned after each other and the web browser will continuously repeat the animation cycle.

Slightly more interactive animations are those to be played by mediaplayers, in AVI, MPEG or MOV format. Plug-ins to the web browser define the interaction options, which are often limited to simple pause, backward and forward. The animation does not use any specific web-environment parameters and have equal functionality in the desktop-environment.

Animations created via VRML or QuickTime VR allow the incorporation of links and become a more interactive 'clickable animation'. These links normally refer to other locations on the Web or start another animation with more or less detail.

In this context it is interesting to mention that at the beginning of 1998 the VRML Architecture Review Board has accepted the GeoVRML Working Group Proposal. This should lead to the incorporation of spatial functionality in future VRML standards. For some animations such as a flythrough, VRML offers interactivity to the user, e.g. the user can define the flight path, and make decisions on direction, during the flight.

6.3 MAPS AND THE WEB: USER PERSPECTIVE

In the previous section some cartographic and technical options have been described but how do these work in practice and how do we access data? 'Access' in this context means finding and retrieving.

Finding something useful on the Web is not a easy task, especially if you are unaware of the exact location. Locations on the Web are given in 'web co-ordinates', the URL address. URLs are dynamic because they tend to change regularly and new ones appear daily. Using the available search engines might give some results, but you can never be sure if these results include all the answers to your query. Some cartographers try to maintain lists of bookmarks referring to organisations dealing with spatial data. A good example is Oddens' Bookmarks (Oddens and van der Krogt, 1996). By September 1998 this list contained over 5500 categorised pointers to map-related web addresses. Although it can be a pleasant pastime to browse through such lists, there is no guarantee that you would find all the addresses of national mapping organisations, for instance. Unfortunately, maps do not play a role in using these lists, although they could well do so.

Another type of initiative to structure spatial data is the Alexandria Digital Library (ADL, 1998). This experimental project at the University of California (Santa Barbara) is an effort to set up a 'distributed digital library' for geo-referenced materials. It aims to give access to digital maps, texts, photographs and other geo-documents. The library aims to be able to answer questions such as: What information is available on this phenomenon, at these locations? Access to the library is via a map-oriented, graphical user interface.

Currently, the most generic and structured initiatives are those linked to National Spatial Data Infrastructures (NSDIs), which themselves are to be linked into a Global Spatial Data Infrastructure. One of the most important parts of such a NSDI is a National Geospatial Data Clearinghouse. This can be considered as a kind of metadata search engine on databases scattered around the country. Organisations, commercial as well as non-commercial, can offer their spatial data products via the clearinghouse. In Europe, this is done according to a description following the European metadata standard. Such standards are of enormous importance in this environment, and indeed they give the user an idea of what to expect from the data they might eventually retrieve. To find data of interest, the user can indicate a region on a map, click a theme, or combine both types of search criteria. Results can be a list of organisations and data sets meeting the query. Hyperlinks will allow the user to access the sites of these organisations where conditions for retrieving the data can be found. Some organisations offer 'teasers', that is test data sets to allow the user to judge the fitness-for-use.

In finding data, the map plays a specific role. For users who want to explore spatial data, the map is a natural access medium or interface. It can structure the individual geo-referenced data components with respect to each other and function as an index. The map will also allow the user to navigate a way through the data.; it is not just a clickable map providing information on data quality for a particular area, but fully hyperlinked, thereby guiding the user to the data set sought. It can be used to connect metadata onto data available for a particular region and, as such, it can also be used to determine the consistency of the database.

When the data have been located, the question remains whether one should download it to explore it locally, or use remote solutions. Each solution will put a certain strain on resources, either at the user's site (client) or at the organisation's site (server) providing the data set. The possibilities will depend on the user's browser, for instance: can it handle plug-ins or work with applets, or does it depend on how an organisation offers its data? Plewe (1997) describes all the current options in a systematic way and gives an extensive list of examples. Downloading full data sets using the ftp protocol allows users to manipulate data in their own GIS environment, but in identifying candidate data sets the user must rely on a metadata description and will only see the data after a full download. Another option is that an organisation has fully prepared data sets available and provides them as bitmaps. These static maps are view only. They can be used as background images in the exploratory visualisation process. There are several alternatives lying between these two extremes. In some situations, the user can define the map contents of a displayed sample. Parameters given by the user determine what will be drawn. Depending on the client-server situation, the user can choose a symbol system or will be given (raw) vector data. Some organisations allow the user to select data that will be pre-processed by the organisation before it is sent to the user, while others offer access to their GIS and let the user execute a spatial analysis first. Today, each self-respecting vendor of GIS software has a solution for the Web and spatial data. Some will work with plug-ins while others use JAVA or JAVA-Script. Limp (1997) compares products from Autodesk, ESRI, Intergraph and MapInfo. Each of these products has its own specific way to deal with graphics, interactivity (spatial analysis), nature of the data behind the graphics etc. For instance, ESRI's ArcExplorer, can be used for downloading, viewing and examining spatial data.

Figure 6.2 Homepage of the Alexandria Digital Library

6.4 MAPS AND SOCIETY: AN EXAMPLE

Morrison's (1997) democratisation of cartography will definitely change our view on maps and map making, such that maps will become individualised products. People are not always interested in very accurate maps; as long as the maps fit their purpose, the user will be satisfied. Although individuals have easier access to data and will be creating more maps themselves, an important role will still remain for the geoinformatics expert. If not in creating communicative maps, then in facilitating how to do so, or in providing tools for exploratory cartography, but also to inform them about new possibilities such as the Web.

An interesting environment to facilitate this transfer of knowledge would be the National Geospatial Data Clearinghouse. In the Netherlands, as in many other countries, the clearinghouse is aiming to become a geoplaza, a sort of shopping mall for all the available spatial data. Most clearinghouses will stop providing service at this level. Their web pages should allow the user to select a location and a theme, and be given an overview of the data on offer. In the current design, maps could play a more prominent role than they do at the moment, as discussed in the earlier section on data access. They could be part of the search engine that tries to find the data, or act as a display of the available data.

However, to add extra value to the clearinghouse for both the user and for participating organisations, it would be a challenge to combine the clearinghouse concept with the concept of a national atlas. A national atlas offers a detailed view of the physical and social aspects of a country. Between the various participating organisations data exists for most of the topics which would be required. Among the participants in the Netherlands are organisations such as the Statistics Office, the Soil Survey, the Geological Survey, the National Planning Office and the Environmental Agency.

The proposition is that each organisation makes up-to-date data available at a certain level of aggregation. One can imagine that the National Survey offers the data for a country's base map, while the other organisations offer data on the themes they represent. Since the idea is that this information is offered free, the level of detail will be limited. In the case of the Netherlands, data could be offered down to a scale 1:1,000,000. More detailed information can be obtained from the respective organisations and should be accessible via links in the atlas or pointers at the geoplaza, the clearinghouse's web page, with links to all participating spatial data providers. An example would be population data, presented on a map with the municipalities as administrative units. On this map, the user can find the most current data of each municipality. However, if users want information on, for instance, the age structure of the population, they are linked through to the Statistics Office or to the particular municipality where the data can be found. The policy of these organisations will determine whether the data are available free or if commercial conditions exist. These sites could provide access to additional data and indeed 'knowledge'. Different exploitation strategies can be designed to realise this vision.

Ideally, the on-line atlas would present data in communicative maps designed by professional cartographers, while experts from the organisation offering the data could write a short description, explaining the maps' spatial patterns. Users could freely download the maps and/or the data for their own purposes. Both the map and the description would provide an opportunity to hyperlink the user to additional information on the mapping method applied, or to the organisation's web site. This approach has an analogy with the current second paper edition of the Netherlands National atlas, which contains maps, text and other illustrations.

However, from a practical point of view, it would be better if the maps could be directly generated by the participating organisations. This is also valid for the meta-information to be used by the clearinghouse. It would guarantee both up-to-date data and maps. With respect to the maps, some design rules must apply since they must all form part of a single on-line national atlas and, as such, some coherence with other maps in the atlas is required. It would be interesting to see if the design rules could be part of, or derived from, the metadata needed by the clearing house to avoid double effort in creating and maintaining this information by the participating organisation. Participating in the on-line national atlas is clearly an additional method to expose the organisation's data to a wide audience. The question is how to convince them it is effective and worth additional effort! Currently, in the Netherlands, a pilot project, initiated by the Netherlands Cartographic Society, is being prepared, involving the Clearinghouse, the National Atlas Foundation and the Royal Library and the Soil Survey.

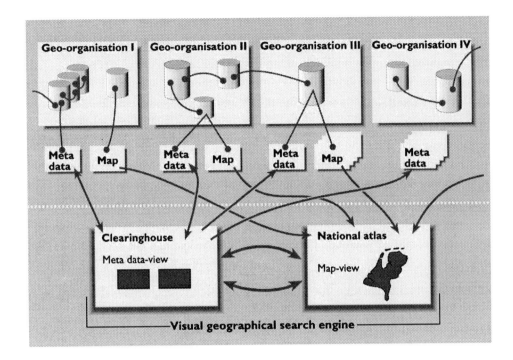

Figure 6.3 Relation between the participating organisations, clearinghouse and national atlas

Visitors might expect to see a consistent national atlas looking the same in whichever medium; paper or screen. However, it is very likely that the visitors to an on-line national atlas would have a different attitude to the imagery they see on-screen than those more traditional users who are used to more traditional national atlases, whether on paper or CD-ROM. The new users are probably eager to click on the maps, and expect action. Not only should the data be free, it has to be interactive as well!

This means presenting the maps as view-only bitmaps is probably not the solution. Although from a commercial point-of-view it might be, since if customers want more information on the maps they have to go to the particular organisation and potentially pay money. However, the maps in a real on-line national atlas should be of the indirect or direct interaction type, as described in the previous section. The indirect maps are used, for instance, to create a base map with the necessary data layers, and the direct maps are used to get to the data behind the map or to go other sources of data or information.

Interaction is required if we wish to consider exploratory cartography. For many users a national atlas is a source of interesting information to browse. Among the functions mentioned earlier, display and query can be realised. However, for some functions, such as re-expression, the users will need the data in their own environment or the data provider will have to offer extensive GIS functionality from their site. From an organisational point of view, this advanced approach might complicate the set up of a national atlas. Although, technically it would be possible, and some examples are known, bandwidth is currently a major problem for this type of interaction.

As such, a national atlas would offer an up-to-date overview of the country's characteristics and provide an extra aspect to the National Spatial Data Infrastructure. This idea could be implemented on both an international, as well as a more regional, scale.

6.5 CONCLUSION

This chapter began with the statement that cartography is changing: changing from being supply-driven to demand-driven. A change that has been driven by what has been called the 'democratisation of cartography'. The role of the Web in this change was discussed, since it also, among other technological developments, stimulates the emergence of a new 'exploratory cartography'. This chapter has provided an overview of the options and limitations for maps on the Web from a cartographic and a technological perspective.

The Web is also an important vehicle for a National Spatial Data Infrastructure and, its component, the Spatial Data Clearinghouses. Such clearinghouses can be combined with the concept of a national atlas to create an up-to-date view of a particular nation. In the Netherlands, research has started to create an on-line national atlas. As well as addressing organisational problems, this research needs to address suitable geographic search methods, and an appropriate interface.

REFERENCES

ADL, 1998, Alexandria Digital Library, http://alexandria.sdc.ucsb.edu/

Harrower, M., Keller, C.P. and Hocking, D., 1997, Cartography on the Internet: thoughts and preliminary user survey. *Cartographic perspective*, Vol. 26, pp. 27-37.

Jiang, B. and Ormeling, F.J., 1997, Cypermap: the map for cyberspace. *The Cartographic Journal*, Vol. 34(2), pp.111-116.

Johnson, D. and Gluck, M., 1997, Geographic information retrieval and the World Wide Web: a match made in electronic space. *Cartographic perspective*, Vol. 26, pp 13-26.

Kraak, M. J. and Ormeling, F.J., 1996, *Cartography, the visualisation of spatial data*, (London: Addison Wesley Longman).

Limp, F., 1997, Weave maps across the Web. *GIS World*, September, pp. 46-55.

Longley, P., Goodchild, M., Maguire, D. M., and Rhind, D. (eds) (1998) *Geographical information systems: principles, techniques, management, and applications*, (New York: Wiley & Son).

Morrison, J.L., 1997, Topographic mapping for the twenty first century. In: *Framework of the world*, Rhind, D. (ed) (Cambridge: Geoinformation International).

Oddens R. P. and van der Krogt, P., (1996) Oddens' bookmarks, http://kartoserver.frw. ruu.nl/html/staff/oddens/oddens.htm

Peterson, M. P., 1997, Cartography and the Internet. Introduction and research agenda. *Cartographic perspective*, Vol. 26, pp 3-12.

Plewe, B., 1997, *GIS Online: information retrieval, mapping and the Internet*, (Santa Fe: OnWord Press).

Robinson A. H., Morrison, J. L., Muehrcke, P. C., Kimerling, A. J., Guptill, S. C., 1995, *Elements of cartography* (6th edition) (New York: J. Wiley & Son).

Whitener, A. and Creath, B., 1997, *Mapping with Microsoft Office, using maps in everyday office operations*, (Santa Fe: OnWord Press).

APPENDIX

Sample sites with bitmap-maps
National Library of Scotland: http://www.nls.uk/collections/maphigh.html
Dutch city maps of Blaeu: http://odur.let.rug.nl/~welling/maps/blaeu.html
Magellan: http://www.pathfinder.com/travel/maps/index.html

Sample site of indirect interaction with map image
Pennsylvania State University DCW: http://www.maproom.psu.edu/dcw/
Multimedia maps of Britain: http://www.city.net/
Projection site: http://life.csu.edu.au/cgi-bin/gis/Map
San Franciso Bay Area REGIS: http://www.regis.berkeley.edu/gldev/regis.html
US Census TIGER files: http://tiger.census.gov/cgi-bin/mapsurfer

Sample site of direct interaction with the map image (clickable maps)
Whisky map of Scotland: http://www.dcs.ed.ac.uk/home/jhb/whisky/scotland.html
Exite travel: http://www.city.net/
Denmark: http://sunsite.auc.dk/Denmark/

Sample site of direct interaction with the map image (clickable maps)
SF 3D VRML: http://www.planet9.com/earth/sf/index.htm
For many links to sample see:
http://www.cgrer.uiowa.edu/servers/servers_references.html#interact-anim

7

A decision support demonstrator for abiotic damage to trees, using a WWW interface

Roger Dunham, David R. Miller and Marianne L. Broadgate

7.1 INTRODUCTION

Forest damage by wind, snow and fire in Europe constitutes the most serious economic problem facing the forestry sector. Wind damaged 100 million m^3 of trees in one night in 1990. In addition, approximately 4 million m^3 are damaged by snow and around 500 000 ha of forest are destroyed by fire annually in Europe (Valinger and Fridman, 1997).

The STORMS (Silvicultural Techniques Offering Risk Minimizing Strategies) project (MLURI, 1998) aims to increase the knowledge of the processes involved in forest damage, and to develop a decision support system (DSS) to help managers reduce the risk of the occurrence of such damage.

The development of such a DSS presents a significant scientific and logistical challenge. The system must be able to assess and predict forest damage risk and allow forest managers to test the effect of alternative management scenarios to reduce the overall risk in both space and time (Hunt and Jones, 1995; Quine *et al.*, 1995). This will allow forest managers to deploy resources strategically, prioritise areas of particular concern and use risk estimates in financial planning of their forests.

A DSS of the nature described requires the integration of components at different spatial and management levels, and different climatological regimes within a European framework. Furthermore, whole risk appraisal over a range of damage types is required because silvicultural strategies employed to reduce one source of risk (e.g. snow) may increase the vulnerability of a stand to other risks (e.g. wind damage) (Quine, 1995).

This chapter addresses generic issues associated with coupling spatial and aspatial models related to damage by wind, snow and fire, to produce a framework for the demonstration of the use and limitations of such models. The modelling tools that have been brought together into a single point of access, operate at different levels of spatial and temporal organisation in a framework structured within a World Wide Web (WWW) interface. These issues include the description and documentation of data and models,

79

using metadata protocols; the incorporation of error and uncertainty in both the modelling and presentation of results; and the structure of a flexible framework capable of extension.

7.2 FRAMEWORK REQUIREMENTS

The framework is required to provide a basis for improved decision making for the management of forests, in which both spatial and aspatial information is used and with which the risks of damage due to wind, snow and fire associated with alternative management scenarios may be explored.

Furthermore, the framework must allow for the facts that: each model would use some spatial analysis functions; each model required custom written code (although ranging widely in degree of sophistication); and, each would be developed by the original modeller.

Four approaches were available for structuring and developing the overall framework: a web-based, hypertext system; a Geographic Information System (GIS), with embedded macros; a single package with bespoke software and with routines for each model; or stand-alone packages for each model. Each of the GIS, single package and stand-alone options has been used before for similar exercises and thus the benefits and potential of each can be defended (Vckovski and Bucher, 1997; Leavesley *et al.*, 1996; Fedra, 1996).

The option of a web-based structure was chosen because it offered the greatest flexibility for independent model development and testing, and facilitated the updating and refinement of individual models. It also enabled the incorporation of a disparate array of contents for the presentation and operation of models, the supporting research and the dissemination of the project content and results, in a way that can be updated centrally, yet permit world-wide access. The approach also has the advantage of using widely available, standard and low cost computing facilities and permitting direct email access to the modeller for discussing future operational use of the models.

The key contents that the web framework was designed to offer were:

1. An interface to explore component models in terms of user requirements (but not to provide data or actually run the models);
2. A package, or set of packages, which demonstrate the use of selected integrated models for the management of damage risk;
3. Information about the issues of scale and error in the models (i.e. quality of model output);
4. Documentation of models and data used in the project.

The virtues of adopting the use of WWW protocols as the basis of the framework are:

1. The ease of designing a common means of presentation;
2. A modular structure which is simple to build upon by connecting to relevant, independent research and models, either deepening or broadening the content of the model demonstrator;
3. Ease of access by local or international users, employing standard, low cost, computing equipment;

4. A rapid and streamlined approach to incorporating model revisions and updated management advice;
5. Model confidentiality can be maintained by the use of CGI scripts or indirect presentation of results using intermediate lookup tables.

The disadvantages are:

1. Currently, under-developed for on-line use of GIS (although improving rapidly);
2. If inadequate security arrangements are in place, then demonstrator users may be able to directly access numerical values which they contain, such as those in equation variables and constants. This can be undesirable in many circumstances, for example, in the case of unpublished (in a scientific sense) data or models;
3. The variation in functionality of options available to web site users depending upon the hardware and software that they are using. For example, some of the facilities provided for viewing data and results using video sequences are not widely available for user of Unix (or similar) operating systems;
4. The access to a remote site can be severely hindered by the communications infra-structure, if the data being transferred to the user is sufficiently large compared to the transfer rates possible and the number of other users either accessing the site or also transferring data at the same time.

7.3 METADATA AND METAMODEL INFORMATION

Underlying the framework, there are several models and other supporting research that have been developed separately. The models have been surveyed using a standardised approach to describing the data (NSDI, 1996; Rhind, 1997) by designing and implementing the collection of metadata and metamodel information, to identify the key links that are required either to couple them together; or, compare their data requirements to those available to the user. Thus, users can test whether the data they have is sufficient for use of the model relevant to their geographical location or circumstances.

The metadata format has been drawn-up to capture the key aspects of the data, including details of its lineage, accuracy, content and ownership, in addition to information about its structure and format (Lanter, 1990). The metadata collection was implemented using forms on the Web which were completed for the datasets for each area and then compiled by the data co-ordinator.

A similar approach was adopted for the collation of information about the component models. A total of 17 models used in the assessment of risk of trees to wind, snow or fire damage were derived. These ranged from logistical estimates of risk based upon 3000 sample plots (in Sweden), to process-based mechanistic models of tree response to wind and snow loadings (United Kingdom and Finland) to spatial models of either wind damage (United Kingdom) or fire spread (in Portugal). The metamodel information consisted of details of model derivation (such as source scales and data); validity and sensitivity testing (providing a qualitative or quantitative assessment of the sensitivity of the models to each input); version and release details and points of contact for the modeller. The collation of the metamodel information was also implemented using web forms. Examples of the metamodel forms can be seen on the web site.

As many of the models are currently unpublished, or have a commercial value, the metadata is not publicly available on the Web, and is subject to password protection. The meta-information has, however, been used in the creation of a page that identifies which models are available for use by a specific user. The user can specify up to 37 different datasets, under topics of data type (DEM, fuel map, etc.), tree characteristics, site characteristics, stand characteristics and climate. The user is then provided with a table of all of the models against all of the input variables. This table highlights those data that the user has available and thus which models may be run. At this stage of the demonstration little guidance is given on the suitability of any particular model with respect to the source scales or coverage of the data. Further guidance is provided at the stage of model demonstration.

7.4 MODEL DEMONSTRATION

7.4.1 Framework design and method of presentation

Table 7.1 summarises the nature of the models which form inputs to the framework, the models developed and the nature of the coupling to GIS that was required by the project and its components. They include mechanistic, empirical, logistic and rule-based models and they range from studies of individual trees, satellite pixels of 30 m spatial resolution to monitoring plots and forest sub-compartments.

GIS software was also used in the pre-processing of the spatially referenced data to produce, or combine, inputs as follows:

- FireMap - height, slope, aspect, representation of wind direction, fuel map classes;
- GALES - height, topographic exposure, soils, management compartments, stand characteristics;
- HWIND - management compartments, proximity to stand edge, stand characteristics.

Additional spatial analysis was undertaken in studying stand variability and the modeling of snow cover but neither dataset has, to date, been used in the operation of the models. The Arc/Info software that was used in each case could be linked directly to the framework in the future, using the purpose written macros, where a PC version of the package was available

The models are presented within the framework of the demonstrator in three ways:

1. Interactive access to single tree or stand level models, with either direct computation of model results or the use of look-up tables. The user can select from up to eight different model inputs and several numerical or thematic values for each variable as appropriate. The interaction is via a graphical interface for operation at national levels for each of Sweden, Finland and the United Kingdom;
2. Illustrative models, of the forest and regional models for Portugal, Finland and the United Kingdom, which provide pre-processed graphical outputs based on the options selected by the user;
3. Direct access to three models, all of which are PC-based. Two of the models operate directly on Idrisi format data files. In each case, the GIS has provided the format and

some of the computation routines for use with spatial data. One package - Firemap (Vasconcelos *et al.*, 1994) from Portugal - links the web demonstrator to an interface written in Delphi (Borland, 1996), into which the fire risk model is written, underpinned by the file operations using Idrisi.

Table 7.1 Summary of component models and their implementation

Level	Model	Damage Agent	Type	GIS Coupling
Tree	HWIND	Wind and snow	Mechanistic	-
	GALES	Wind and snow	Empirical	-
Forest/	HWIND	Wind and snow	Mechanistic	Loose coupling (Arc/Info)
Stand	GALES	Wind and snow	Empirical	Optional close coupling (Idrisi)
	Valinger	Wind/snow	Logistic	-
Regional/	FireMap	Fire	Rule-based	Optional close coupling (Idrisi)
National	Valinger	Wind/snow	Logistic	-

(Notes: 1. HWIND and GALES operate with an optional snow loading, but the Valinger model combines the damage agents into the risk models. 2. There are 14 Valinger models which operate on different species and in different regions. The operation of HWIND and GALES at a forest stand level is via a spatial allocation of input variables.)

7.4.2 Interactive models

To initiate the interactive demonstration of the models relating to wind and snow damage, the user is presented with a choice of tree species - Scots pine (*Pinus sylvestris*), Norway spruce (*Picea abies*), Sitka spruce (*Picea sitchensis*) and birch (*Betula* spp.) - and country, or region, of interest (Finland, United Kingdom and north-, mid- or south Sweden). Only those combinations of country and tree species for which the models have been tested are made available.

The interface uses a standard format such as that shown in Figure 7.1 which consists of:

1. a map of the country on the left of the content area (in this example Sweden);
2. a button bar for selection of Instructions for use of the demonstration; notes on the model and its data, including access to the metadata and metamodel forms; and an email link to the model developer;
3. options for the selection of inputs to the model variables such as soil type, tree species, tree height, crown length, altitude and location (via pointer at the country map);
4. the output of estimated risk of damage presented to the user, together with an estimate of the risk with a 20% error in the tree height data.

The standardised approach to coding and presentation illustrated for mid-Sweden in Figure 7.1 has also been adopted for each of the wind and snow models operating in Finland and the United Kingdom. In each case, the model inputs and outputs are slightly different from those required by the model presented for Sweden, with the outputs being the wind speeds required for over-turning and stem-breaking, and the inputs depending upon those factors which are important determinants of wind and snow risk in each country.

7.4.3 Illustrative models

The presentation of the inputs and results of modelling is presented for study forests in the United Kingdom, Portugal and Finland. In each case, the user can view maps, or images, of the inputs to the models for operation at a forest or stand level. The views of the model outputs are illustrations of pre-selected input combinations, showing how the spatial pattern of inputs and outputs varies across each study site.

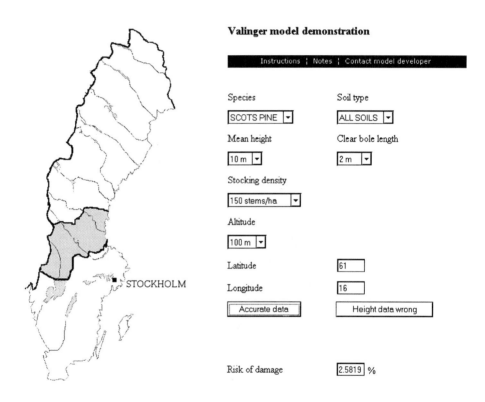

Figure 7.1 Example of interface for use of logistic model developed to estimate the risk of forest damage in Sweden. This example provides information about Scots in mid-Sweden, but other models have been developed for other parts of the country and for Norway spruce and birch. The user can change the parameters shown, and the location of the tree. 'Accurate data' returns the risk of damage if the tree measurements are correct, 'Height data wrong' returns the damage risk if height is inaccurate by 20%.

7.4.3.1 Finland

The content of the Finnish forest demonstration includes a detailed description of the working of the component aspatial models (HWIND and Valinger risk model). It also includes spatial modelling (HWIND and Valinger models coupled to the Arc/Info GIS) as applied to the assessment of the risk of damage at a forest stand level.

7.4.3.2   *Portugal*

The Portuguese study area illustrates the use of FireMap which uses the Idrisi GIS as the basis for its spatial functionality. Detailed descriptions of the model and the results of the experimentation are presented within the WWW framework under the 'Supporting Research'.

   To demonstrate the sensitivity of an area to fire spread with respect to different wind directions and speeds, temperatures and humidity, the user is able to select from these variables and obtain a map of the risk of fire (in four classes) and a statistical breakdown of the proportions of the study area ascribed to each risk class. An interpretation of the results is also provided to aid the user in assessing the risk of fire within the area and the likely flame lengths and fire line intensity. Finally, the user is provided with management approaches to fire suppression for each level of risk and thus the user can compare the viability of the suppression techniques against the contributing causes of fire risk and changes in risk.

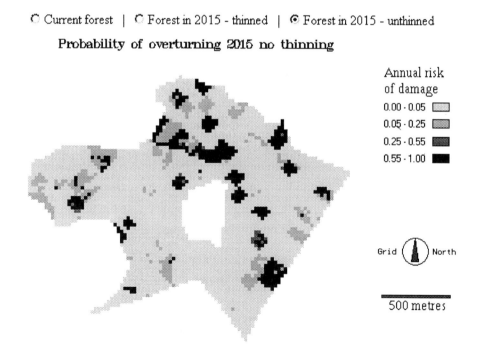

Figure 7.2   Example output for Cwm Berwyn forest in mid-Wales (SN 740565). The example indicates the calculated risk of damage for regions of the forest in the year 2015 if a non-thinning regime is followed until that date. The WWW demonstration also indicates the current risk of damage and estimated risk of damage in 2015 if a thinning regime is followed instead.

7.4.3.3   United Kingdom

The demonstration of the study forest presented for the United Kingdom (the Cwm Berwyn forest in Wales) has included scenarios of the impact of different management decisions on the levels of risk of tree-overturning and tree-breakage. The scenarios are the thinning or non-thinning of the forest and the outputs are shown for the risk levels in 2015 (when the trees will be at their maximum mean annual increment), together with the risk estimates when errors in tree height of 20% are included. An example output is presented in Figure 7.2.

7.4.4   Direct access to the full models

A version of the modelling software that permits direct access to the GIS macros and the full working models (depending upon the computing configuration) has been linked to a specific WWW page. While this considerably increases the power of the demonstration this option is not publicly available due to concerns regarding the security of copyright.

7.5   ERROR AND UNCERTAINTY

7.5.1   Introduction

Information on the certainty and accuracy, and their propagation in the outputs of models is of significance to any user (Veregin, 1995; Goodchild, 1996). The sources of uncertainty include the data (spatial or non-spatial), the algorithms used in deriving non-measured model inputs and a further factor is the geographic scale of the operation of the model. The demonstrator presents information to the user in two ways:

1.   Examples of the consequences of error in each of the sources are presented in relation to an agent of abiotic damage and discussed in terms of the limitations of the applicability of the models.
2.   Interactive demonstrations of the models include the presentation of numerical estimates of risk with respect to a variable to which the model is sensitive (such as tree height, soil classification or fire fuel type).

Three specific examples from the sources of error and uncertainty are described below.

7.5.2   Resolution of DEM data

Two examples were used to illustrate the impact of resolution on derivations from DEMs: the calculation of topographic exposure and the measurement of surface variability.
      The DAMS model of topographic exposure (Quine and White 1993) was calculated for eight different resolutions of DEM: 2 m, 5 m, 10 m, 20 m, 40 m, 50 m, 100 m, and 200 m. The DEM data was derived from 1:10 000 scale maps for a 5 km x 4 km test area in Fetteresso in Scotland.

The eight different resolutions of DEM that were used as inputs to the DAMS scoring system were compared cell-by-cell, at the resolution of each cell. On average the DAMS value was 4 units higher for the 2 m cells than for the 50 m cell size, with the greatest differences being in the steeper valleys, a reflection of the ability of a higher resolution DEM to better represent steeper slopes. These results have been derived for just a single study site and therefore, the typicality of the magnitude of the differences in different topographic conditions is not tested. However, the method can be used to assess the stability of the scoring system with respect to topography.

The second example of the effect of data resolution on uncertainty compares the surface variability of eight DEM resolutions derived by digital photogrammetry for a forest in mid-Wales (Cwm Berwyn). As described by Miller *et al.* (1997), a DEM produced in this way can be used in the measurement of tree height and canopy variation. The DEM data were compared on a cell-by-cell basis and by measurement of the local standard deviation of the DEM surface. Evidence of the canopy as a distinct feature (as indicated by the values for the local standard deviation) begins to be apparent at 50 m resolution. However, it is not until resolutions of greater than 10 m that the estimates of tree height are within a mean of ±3 m. At higher resolutions, the variability in the canopy with respect to underlying terrain features, such as drainage patterns are evident and the mean accuracy of the estimation of tree height is better than ±1.5 m. The results of this study can be used to guide the user in the selection of DEM resolution for assessing canopy height remotely and using this as an input to measures of canopy variability.

7.5.3 Classification and scale of soils data

When an area of variable soils is mapped, some of the mapping units produced may contain more than one soil type, because it is not practical to map small inclusions of soil separately in a polygonal dataset (Fisher, 1992). This issue of inclusions arises irrespective of the scale of the map or source data (ground survey or aerial photographic interpretation).

Many of the models developed in this project use soil type as an input. Errors associated with mapping of soils will, therefore, propagate through the model resulting in errors in the risk estimation. To illustrate the magnitude of the problem of classification, a 1:50 000 scale soil map (MISR, 1980) was digitised for a forest in the West of Scotland (Cowal). The soils were recoded according to the scheme used by the GALES model using four, different, interpretations of the soil key: main soil class within the polygon; 'inclusions' soil class (based on sub-class information); and the best and a worse case scenarios based on which was the most or least stable soil type of the 'main' or the 'inclusions' soil code. Maps of the main soil type at scales of 1:10,000 and 1:1,000,000 for the same area were also investigated.

For the 1:50,000 scale, the best case scenario suggested that 93% of the area consisted of 'stable' soils, whereas in the worst case scenario only 25% of the area was occupied by such soil types. This was reflected in large changes in the calculated area of forest at risk of damage from wind and snow depending on which method and scale of soil classification was used, and outputs are provided within the demonstrator. Examples of the model outputs based on the best case (i.e. most stable) and worst case (least stable) soil scenarios are presented in Figure 7.3.

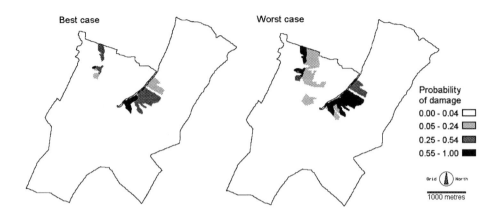

Figure 7.3 Annual risk of wind and snow damaged calculated by forest GALES model for two soil classifications based on 1:50,000 soils map for a forest on the Cowal Peninsula (NS 115930). Each polygon of the soils map has a main soil type, but sub-class inclusions may also be recorded, and may represent a soil type with very different properties from the main soil type. The left image indicates the 'best case' scenario with the risk calculated according to whichever soil type (main soil type or inclusions) within a polygon gives the tree most support. In contrast the right image illustrates the 'worst case' scenario.

7.5.4 Algorithms used for deriving data.

Risk of fire spread is highly dependent on slope, so variation in slope will cause variation in fire risk estimation. The scoring of 'risk' associated with individual pixels will therefore vary depending upon the slope algorithm used, but the significance of this will only be determined by the weighting of the slope images within the FireMap processing.

Algorithms as a source of error in models were assessed and illustrated by reference to the derivation of slope and aspect from DEMs. A test of eight different algorithms described in Jones (1998) was undertaken for deriving slope and aspect as a local property of the square lattice/grid type DEM using a Portuguese test site as a example. The impact of the selection of different algorithms was summarised in both statistical and spatial terms with respect to the calculated areas of each fire risk class.

In comparison to Fleming and Hoffer's method, the other algorithms produced risk maps where on average 4.7% of the cells differed by at least one risk class. This ranged from 2.2% for the Diagonal Ritter's method to 6.3% for the simple method. The average down-dip method the maximum downward gradient method and the simple method all resulted in some pixels changing by two risk classes whereas all other algorithms resulted in changes of just one class.

7.6 DISCUSSION

The decision support demonstrator described in this chapter makes efficient use of the Web to illustrate the models that have been developed within the STORMS project. Using the Web has the key benefit of allowing individual parts of the demonstrator to be altered or upgraded without requiring the recompilation of the entire interface.

Currently the Web is still underdeveloped for using GIS based models 'on the fly', and while this problem is being resolved as technology advances, other issues must be addressed.

1. Allowing 'live' access to the modelling tools implies that either the models are held centrally, in which case data must be transferred across the Internet, or else the models themselves will be transferred allowing users to access them on their local machine. While dissemination of models would allow their use in situations with a slow or non-existent network, it also has two major disadvantages. Firstly, it exposes the models to copying, modification and reverse engineering which has implications for models which have a financial value, since this value may be reduced or lost. Secondly, if models are disseminated then users may not 'upgrade' to subsequent versions of the model, so that outputs may be unreliable. In contrast, if models are held centrally, and the data is transferred across the Internet, then updated models can be 'plugged in' to the demonstrator as they are developed, ensuring that only the most up to date versions are being used. Furthermore, centrally held models will be more protected than distributed models from copyright infringement. On the other hand, the transfer of large volumes of data across the Internet for processing may be relatively slow if the network bandwidth is not sufficient for the volume of data being transferred. There may also be a reluctance of data owners to transfer data outwith their direct control for processing.

2. The reliability of both models and data is the most critical issue associated with the aiding of decision making. Several possible sources of error have been discussed with reference to this demonstrator, and indications of the magnitude of uncertainty in model outputs have been presented. The consequences of error and uncertainty for an operational user could be reduced by allowing access to the full meta-data and meta-model documentation, but this would also compromise the intellectual copyright of commercially valuable models. An alternative, but less elegant, solution to this problem is not to allow users direct access to the models at all, but, instead, to require them to contact the model developers (which can be achieved automatically within the demonstrator via e-mail) to discuss their requirements.

3. Copyright issues associated with model outputs must also be addressed. In recent years legislation has been introduced in both Europe and the United States to address the issues of the availability of digital data, copyright over its use and products derived from it and the costs of such data. Organisations that are suppliers of potentially suitable data, whether they be national agencies or commercial companies, are still evolving their policies to ownership, copyright, access and more recently the potential of a legal liability for the use (or misuse) of their data.

The generic issues associated with the role and use of the Web in the operation of models have been discussed and exemplified through its use in supporting decisions on the risks of abiotic damage to forestry in parts of Europe. These generic issues included the accessibility of models, or the modeller, to the user; the protocols for documenting data and models; the presentation of the sensitivity of outputs to errors in either data or models; the extent to which the user may include their own data in the model demonstrations; and the efficiency and practicality of operating models over the Internet.

In conclusion, the chapter has presented a description and discussion of the key issues associated with an approach to integrating different types of models, tackling different physical problems at different scales. The implications for decision makers and potential model users of the opportunities and difficulties when wishing to access and use a comprehensive package of modelling tools for quantifying the risk of damage to forests are discussed and some guidelines identified with respect to model applicability.

ACKNOWLEDGEMENTS

The authors wish to acknowledge funding for this research from the EU project AIR3-CT94-2392 and from the Scottish Office Agriculture, Environment and Fisheries Department. Contributions from other project participants are also acknowledged, these are: University of Joensuu, Swedish University of Agricultural Sciences, University of Aberdeen, CNIG (Portugal), Forestry Commission, University College Galway, Direcçao Geral des Florestas.

REFERENCES

Borland, 1996, *Delphi for Windows 95 and Windows NT: Users guide*, (Scott's Valley: Borland International).

Fedra, K., 1996, Distributed models and embedded GIS: Integration strategies and case studies. *GIS and Environmental Modeling: Progress and Research Issues,* (Fort Collins: GIS World Books), pp. 413–418.

Fisher, P. F., 1992, Real-time randomization for the visualization of uncertain spatial information. In *Proceedings of 5th International Symposium on Spatial Data Handling,* (Charleston: IGU Commission on GIS), pp. 491–494.

Goodchild, M. F., 1996, The spatial data infrastructure of environmental modeling, *GIS and Environmental Modeling: Progress and Research Issues,* (Fort Collins: GIS World Books), pp. 11–15.

Hunt, H. and Jones, R. K., 1995, Building a forest management decision support system product: challenges and lessons learned. In *Proceedings Decision Support 2001.17th Annual Geographic Information Seminar, Resource Technology '94 Symposium,* (Bethesda, Maryland: American Society for Photogrammetry and Remote Sensing), Vol. 1. pp. 92–102.

Jones, K. H., 1998, A comparison of algorithms used to compute hill slopes as a property of DEM. *Computers and Geosciences,* Vol. 24, pp. 315–323.

Lanter, D. P., 1990, Lineage in GIS: The problem and solution. *National Center for Geographic Information and Analysis: Technical Report* 90-6.

Leavesley, G. H., Restrepo, P. J., Stannard, L. G., Frankoski, L. A. and Sautins, A. M., 1996, MMS: A modeling framework for multidisciplinary research and operational applications. *GIS and Environmental Modeling: Progress and Research Issues,* (Fort Collins: GIS World Books), pp. 155–158.

Miller, D., Quine C., Broadgate, M., 1997, The Application of Digital Photogrammetry for Monitoring Forest Stands In *Proceedings of the International Workshop on Application of Remote Sensing in European Forest Monitoring,* Kennedy, P. J. (ed) (Brussels: European Commission), pp. 57–68.

MISR, 1980, *The soils of western Scotland, 1:250,000,* (Aberdeen: Macaulay Inst. for Soil Research).

MLURI, 1998, http://bamboo.mluri.sari.ac.uk/aair

NSDI, 1996, http://nsdi.usgs.gov/nsdi.html

Quine, C., 1995, Assessing the risk of wind damage to forests: practice and pitfalls. In:. *Trees and Wind,* Coutts, M. P. and Grace, J. (eds) (Cambridge: Cambridge University Press), pp. 379–403.

Quine, C. P. and White, I. M. S., 1993, Revised windiness scores for the windthrow hazard classification: the revised scoring method. *Forestry Commission Research Information Note* 230, (Wrecclesham: Forestry Authority Research Division).

Quine, C., Coutts, M. P., Gardiner, B. A., Pyatt, G., 1995, Forests and wind: management to minimise damage. *Forestry Commission Bulletin 114,* (London: HMSO).

Rhind. D., 1997, Implementing a global geospatial data infrastructure (GGDI), http://www.frw.ruu.nl/eurogi/forum/ggdiwp2b.html.

Valinger, E. and Fridman, J., 1997, Modelling probability of snow and wind damage in Scots pine stands using tree characteristics. *Forest Ecology and Management,* Vol. 97, pp. 215–222.

Vasconcelos, M. J., Pereira, J. M. C. and Zeigler, B. P., 1994, Simulation of fire growth in mountain environments. In: *Mountain Environments and Geographic Information Systems,* Price, M. F. and Heywood, D. I. (eds) (London: Taylor and Francis), pp. 167–186.

Vckovski, A. and Bucher, F., 1997, Virtual datasets – smart data for environmental applications. In: *Integrating Geographic information Systems and Environmental Modelling,* (Santa Barbara: NCGIA).

Veregin, H., 1995, Developing and testing of an error propagation model for GIS overlay operations. *International Journal of Geographical Information Systems,* Vol. 9, pp. 595–619.

8

Integrated Regional Development Support System (IRDSS)

An innovative distributed Geographical Information System, serving Europe by the use of emerging Internet technologies

Adrian Moore, Gerard Parr, Donna McCormick and Stephen McAlister

8.1 INTRODUCTION

As part of the European Union Fourth Framework Telematics Programme the University of Ulster (UU) Telecommunications and Distributed Systems Research Group and the Geographical Information Systems Research Team, have been involved in developing IRDSS (Integrated Regional Development Support System). The project is being developed by a pan-European consortium of organisations led by the European regional authorities organisation ERNACT (European Regions Network for the Application of Communications Technology). The aim of the project is to enable geographically, economically and socially marginalised communities and groups to develop the full potential of their area, through access and sharing of information. As the spread and intensification of global competition becomes increasingly felt at the regional level, regional development strategies must respond by taking advantage of the benefits of new and emerging Information and Telecommunications technologies. For more peripheral, socially and economically deprived regions there is an urgent requirement to connect their economies, to those of more central regions and to the emerging Information Society. At the intra-national level the responsibility for developing such strategies lies in the main with local authorities.

The potential of accessing data and information for economic development and decision-making, via a real time information network, is a relatively new phenomenon, outside organisational Intranets. This potential has been realised over the past few years with the explosion of Internet usage, especially with the World Wide Web (WWW) development and standardisation of web server technology. To date, although the broad economic and social benefits of Geographic Information Systems have been identified, accepted and widely exploited using desktop and intranet technologies, the potential of expanding GIS services and systems via telecommunications technologies is a relatively new concept (Crowder, 1996; vanBrakel and Pienaar, 1997; Cole, 1997 and Heikkila, 1998).

A key objective of the IRDSS project is to utilise the potential of emerging telematic tools to provide a World Wide Web based, platform independent, Geographical Information System serving a comprehensive European Regional Authority database. It provides on-line, user friendly access to integrated, multi-media spatial data on a local, regional and cross-border basis. Data contained within IRDSS include spatial planning, business, tourism, socio-economic and environmental themes. Access to these data themes can be provided via user friendly multimedia kiosks in public areas (such as libraries, community centres), via personal computers in offices (for governmental and local administrators within a region) and to the citizen via an Internet connection.

The aspect of 'where' often transforms data into valuable information, which enhances its usefulness. The importance of Spatial Datasets is evident in today's information society, from defining a consumers market area, to weather forecasting and environmental monitoring. Hence spatial datasets are increasingly being incorporated into systems to help improve the quality of life, especially in regions which have traditionally been excluded from new technology and economic development.

8.2 POTENTIAL BENEFITS OF THE IRDSS INFORMATION

The types of user that are targeted by IRDSS fall into two groups. The first includes local community/development groups, businesses and local administrators, whilst the second includes regional administrators and local government officers. Some of the potential benefits of the system are discussed below.

The Tourism theme contains information on all the tourist and recreational sites, and facilities within a region, including contact details, admission charges and type of facility. Thus it contains current information, similar to that held by each regions Tourist Information Board, such as accommodation, but with the added benefit of geography. This will give the client a more informed choice and enable the user to carry out such queries as 'show all tourist sites and facilities within ten miles of a particular hotel or bed and breakfast that they may be interested in.

IRDSS aims to assist planning policies and promote public access to planning proposals, via the Spatial Planning theme, which contains details of all the planning developments within each region. Thus, users will now have current planning proposals at their fingertips. This will assist both the Planning Departments in their decisions and Strategy Plans and local organisations and environmental bodies to evaluate developments before planning permission is granted.

To promote economic activity, the Business Theme contains a register of all the companies within each region, providing details on such aspects as products, industry type, number employed and business contact details.

Such information is useful from local authority level down to the individual interested in Regional Strategic planning, business marketing and planning. This information is complemented by the ability to integrate socio-economic information such as the census and infrastructural information such as communication networks.

8.3 OTHER GIS INTERNET SOLUTIONS

There is a reasonable increase in the number of GIS based Sites and Information Services available on the World Wide Web. As part of the IRDSS project an extensive and regular appraisal of other web-based solutions was undertaken, to allow IRDSS to benefit and learn from others weaknesses and help develop the best system possible. These Internet GIS/Mapping Systems were found through World Wide Web Search Engines, and examined in the context of the IRDSS Project and in the light of the user needs study (see section 8.3.1).

Appendix 1 contains a list of a number of web-based GIS, and their web addresses (URLs), with a brief description.

8.3.1 Observations on current web-based systems

(a) In some Internet GIS/Mapping Systems numerous windows may appear, running several JAVA applets for maps and table displays, which may present a cluttered interface. This issue will be avoided in IRDSS, by limiting the number of opened windows/applets displayed, to a maximum of four.

(b) A Java applet query ability is very useful and user friendly and IRDSS will incorporate this into its own system.

(c) Providing the user the ability to view relevant metadata is also necessary to allowing informed data comparisons to be undertaken

(d) It is important for IRDSS to present its information in a user-friendly and accessible way (especially for downloading) if IRDSS is to be commercially successful.

(e) The vast amount of data, both in tabular form and as satellite images, need to be presented in defined European standards.

(f) IRDSS needs to be a flexible system, which will allow users to create and save for future use, their own canned queries.

(g) The ability to view data in various forms at any one time is also important (i.e. as a map view, a table and a graph).

From an investigation of GIS Web Solutions, most would not satisfy the needs of the typical IRDSS user. It is therefore safe to conclude that the development of a new IRDSS solution was necessary to meet the 'User Needs' determined by the Local Authorities, to provide a cutting edge solution in the field of Internet GIS, rather than an off-the-shelf software package. There are many more GIS Web Solutions available, but those reviewed covered the spectrum of the typical range of development at the current time (see Appendix 1.)

8.4 THE IRDSS GIS SYSTEM

The IRDSS system:

- is generic and can utilize various propriety GIS systems (currently MapXtreme, MapObjects and ActiveMaps), combining the latest GIS, Internet, client–server and distributed database technologies;
- provides an open-system based software application, that can be integrated with existing in-house data systems, including data from local and regional authorities;
- provides spatial data analysis capabilities, (i.e. the selection of type X industries

which fall within radius Y of site Z and are in areas with more than 20%
unemployment.);
- uses current data standards, with National Cartographic data (both raster and vector
  maps) and multi-media images;
- will use a specially designed, multi-lingual user friendly interface, adaptable for use
  by citizens, local communities and development organisations.

The current distribution of regions involved in IRDSS can be seen in Figure 8.1.

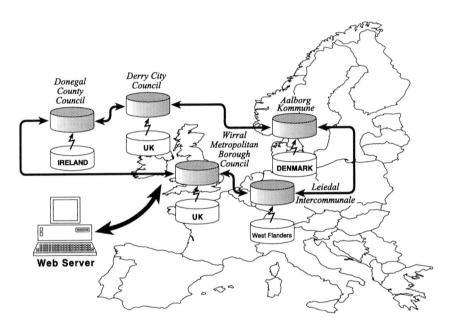

Figure 8.1   The distribution of IRDSS sites

8.4.1 Functionality

The functionality of IRDSS was determined by the requirement to develop a system that
would meet the 'user needs'. To meet this requirement IRDSS would have to provide the
functionality of a GIS and not just be a simple mapping system. It also needed to support
the GIS and data the Local Authorities already had, yet provide a common interface with
the same functionality to facilitate comparison. A summary of the functionality
requirements to meet the original objective includes:

View & Select function requirements
- Inform users about browser recommendations and plug-ins.
- Handle raster and vector data, search results and graphic representation.
- View and select/deselect more than one information layer or theme, at any time.
- Must support the presentation of all types of media.
- Save one or more set of results generated by previous searches for access outside the
  system.

- Enable users to select the language and the region in which they wish to search.
- Link to any necessary plug-ins and their automatic installation.

Search & Analyse function requirements
- Perform specific searches and queries on the information.
- Combine relevant search parameters and values under each theme and region.
- Perform analysis (arithmetic, logical and statistical operations) incorporating more than one theme and region.
- Compare specific information and search results.

Manipulation-function requirements
- Enable users to customise the interface facilities.
- Manipulation of maps and search results and the ability to export these.
- Enable users to delete or print search results.

These functional requirements have been divided into two main sets of tools (Map Manipulation & Data Manipulation Functions), available as icons within IRDSS.

8.5 PROBLEM ISSUES FOR A WEB-BASED EUROPEAN IRDSS

IRDSS involves the integration of Geographical Information Systems (GIS) and the Web to act as a base upon which a number of thematic data sets are overlaid. This novel system fully utilises the comprehensive database capabilities of GIS and spatial data analysis. This not only enables the provision of mapping facilities, but also the ability for on-line access to and visualisation of, datasets and query results. Specifically it is the ability to provide Query Transaction processes that increases the functionality of the system, via the World Wide Web. As a web-based European system, IRDSS encounters a few general problem issues, including those associated with building a common data model for different European regions.

IRDSS contains a common data model for all five themes, even though some regions hold their information in different spatial units, overlaid on vector maps in one region, raster in another. Different regions have different problems with the procurement of data and maps. This data model is derived from information Local Authorities already had or will have in the near future. This makes it relatively cost effective and easy for any other European Local Authority to join in IRDSS and add the information they already contain in their databases, where appropriate.

8.5.1 Security and copyright/price of data

The question of the freedom or charging of information and open government has become more of a problem in recent years with the availability of computer-based information over networks. (Crowder, 1996). With the introduction of distributed GIS, certain data ownership issues arise which were extensively investigated as part of IRDSS, these include data security, copyright, data standards and who pays for data (users or Local Authorities). Consequently it was decided that IRDSS should support both thin and thick client web-based GIS software, and thus the thin client solution is used when viewing Ordnance Survey Map data in UK and Ireland and for information where the data providers are wary of public provision.

These sort of systems need a firewall to prevent unauthorised access of data and potential corruption of the data structures and data items. Protection for software copyright is relatively new, as it has only existed since the mid 1980's (Webster, 1997). IRDSS has development copyright on the maps viewed, and this will also act as a deterrent upon unrestricted screenshots being taken of the system. Census and Ordnance Survey retain the copyright over their data and maps. Various organisations providing data are currently using IRDSS as a test-bed for charging and advertising how their data can add value to data belonging to many other organisations. We are currently investigating how a charging system could be put into operation, probably along the lines of an electronic dedicated charge card i.e. a C-SET, once IRDSS becomes commercial. The C-SET (Chip Secure Electronic Transaction) initiative is one of the ISIS (Information Society Initiatives in support of Standardisation) projects. (GI2000, 1998).

This issue is also under investigation in another major initiative within the University of Ulster (BORDER, 1998)

8.5.2 Data comparability

This is a main issue across all types of map and thematic data, with Census data being a classic example (Waters, 1995). The problem is addressed in IRDSS with the provision of a data-dictionary to explain how each item of information was collected, owned by whom and its accuracy. This will hopefully allow the client to make an informed decision before undertaking data comparisons between different regions. In the near future it is hoped that through the work being undertaken by Eurostat and other organisations, much of the statistical data will become standardised across Europe especially in how Census information is collected and processed.

8.5.3 Standards

Research was undertaken into data standards and display standards in use in Europe and specifically in each of the regions involved in the IRDSS project for adaptation in the IRDSS project. This research has not highlighted universal icon standards in use in Europe but work is ongoing at present in the following three areas:

- ISO/IEC DIS 1158-1 Information Technology -- User Interfaces – Icon Symbols and functions --Part 1: Icons --General
- ISO/IEC DIS 1158-2 Information Technology -- User Interfaces – Icon Symbols and functions --Part 2: Object Icons
- ISO/IEC DIS 1158-3 Information Technology -- User Interfaces – Icon Symbols and functions --Part 3: Pointers

The problem with standardisation is that different styles and sizes of icons with the same meaning might be appropriate for different professions. Artistic material may require more 'flowery' icons than scientific material. Symbol copyright is also a consideration, but note that very few icons have associated copyrights.

Research into tourism symbols shows that most systems presenting tourism information have used a variety of icon and symbol styles. Examples include tourism information systems in Berlin, Brussels and Paris, which use actual photographs as icons.

Irish tourism information systems use a wide variety of photographs and styles of icons. The single standard icon found is the large *i* for tourism information. Our research has not highlighted agreed universal icon standards, in use in Europe, for all the information themes involved in IRDSS. There are very few systems presenting information using multi-media or GIS, which use icons and symbols, which are universal standards.

There are several current European initiatives in standardisation, e.g. ISIS, an industry and market-oriented programme. There are also the following Geographic Information (GI) initiatives; GI in the IMPACT and INFO2000 Programmes of DG XIII/E, GI in the Fifth Framework Programme, GI used by EU Institutions, pan-European GI and the EUROGI initiatives. IRDSS will take on board any European data standards emerging in the near future.

Other issues that were taken account of in IRDSS include the possible unwieldy size of some raster maps (over 40MB) and the subsequent implications on performance and download time. This can be overcome by reducing the map information sent over the system, using the smaller map section viewed by the client at any one time. There are also the issues concerning the use of compression tools to be considered.

8.6 IRDSS – THE GENERIC ARCHITECTURE

The technical architectural design of IRDSS combines the latest geographical information systems, Internet, client-server and distributed database technologies. IRDSS provides an open system based software application that can be integrated with existing in-house systems and data from local and regional authorities.

The IRDSS Architecture takes into consideration the need for IRDSS to access solutions from MapInfo (MapInfo, 1998), ESRI (ESRI, 1998) and ActiveMaps (ActiveMaps, 1998). IRDSS utilises a common Java interface to connect to a local authority's GIS solution. The GIS solution in turn connects to the respective regional database. Currently these databases are stored in MS-Access-97 format and the structure is identical for each region involved in IRDSS.

The design of the IRDSS architecture is therefore 'future proofed' as new GIS solutions could easily be incorporated into the IRDSS model when they become available.

8.6.1 IRDSS – the system

As can be seen from Figure 8.2, access to IRDSS is via a central web server. This provides links to the IRDSS GIS solution on the local IRDSS regional server as well as providing links to other IRDSS regional servers and to general information and help pages.

The World Wide Web by which IRDSS is accessed, is based on a client server architecture. The Web Server holds the IRDSS general information pages that the client accessed initially and the IRDSS menu structure, which controls the user interaction and displays the results. The web browser (one half of the client/server relationship) takes the user's request, accesses the relevant data from the IRDSS Web Server, formulates a response and presents it to the user. The other half of the client/server relationship is the server. On the Web, the server contains documents/data and returns them to the IRDSS client when requested. The web server used by IRDSS (Microsoft Internet Information Server 4), enables the return of text and multimedia documents to browsers and is also capable of executing complex programs.

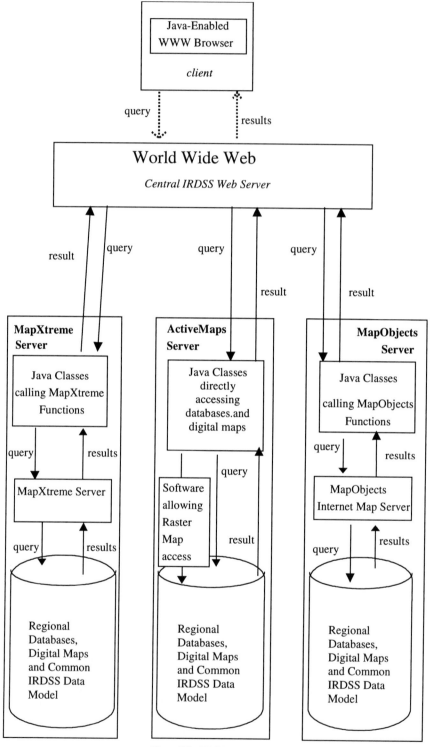

Figure 8.2 IRDSS architecture

Upon entering IRDSS, and transparent to the user, access is provided to either the MapInfo, ESRI, or ActiveMaps solution. Each solution accesses the IRDSS Regional Data Store that contains the data themes on their relevant region and the Common IRDSS Data Model.

The GIS system then provides a result which, is returned to the query originator in the form of a map, thematic map or data result. The optimum system requirements for these technologies were investigated, code was developed for the enhancement of the system and sample applications were prototyped. The system was then subjected to a period of peer review and technical evaluation across Europe.

8.6.2 MapXtreme

MapXtreme is a mapping application server which provides a set of powerful mapping functions to allow data to be spatially manipulated and displayed. It was developed by MapInfo in conjunction with HAHT Software Inc. MapXtreme is used by IRDSS to provide the functionality discussed in Section 8.4. It makes use of MapX, which is a client-side OCX. This technology is platform and browser independent, and results in IRDSS being able to run in any browser, whilst also enabling the IRDSS developed to utilise Microsoft's Active X technology and Java Technology.

8.6.3 MapObjects

MapObjects is a collection of mapping and GIS components which enable developers to add dynamic mapping and analysis capabilities to their applications. It allows applications to be developed using standard Windows development environments such as Visual Basic, Delphi, Visual C++ and Microsoft Access. It is built on Microsoft's Object Linking and Embedding (OLE) 2.0 standard, and is an OLE Control (reusable software component) which comprises 35 programmable OLE automation objects. These can be used like building blocks to create and integrate Windows applications. It allows application builders to use vector data in the form of ESRI Shapefiles or Spatial Database Engine layers, and raster image format such as .bmp, .tif and others.

8.6.4 ActiveMaps

ActiveMaps is an 'Internet Dynamic Mapping System' and it relies heavily on Java to allow users to display and query maps and associated data via the Internet. It interacts with users via Java applets which have been downloaded to the user's computer, and sends messages to the regional web server in order to extract data from databases and to extract maps and boundary files from spatial data files. The map and boundary data is sent to the user in vector format, and this results in faster processing than the alternative which is generating a graphic on the server and sending that to the user for displaying within an HTML page. Thus map data is downloaded to, and manipulated within, the user's computer. Database queries are transmitted to the server and operate directly on the data files.

The current working system can be accessed by emailing the authors.

8.7 CONCLUSION

The System developed by the project team is at the forefront of the exciting integration of the new technologies of Geographical Information Systems, multimedia and Telematics. IRDSS is a developing product and as such demonstrates how new technologies can be applied in a European context to provide enhanced public information access and a wide range of economic benefits to the communities which it serves. Also highlighted are just some of the issues and problems confronting developers in this field, especially at the cross-national level. The dramatic pace of improvements in computing and telecommunications technologies will no doubt facilitate the opportunity to develop and realise the 'distributed GIS' concept more rapidly in the future. Indeed, some of these new technologies are being incorporated into another major project BORDER (Business Opportunities for Regional Development and Economic Regeneration), which has been funded under the EU INTERREG II Programme (BORDER, 1998)

ACKNOWLEDGMENTS

The UU Team would like to acknowledge the wider IRDSS Working Group, including ERNACT (Lifford), Derry City Council, Donegal County Council, Aalborg Kommune, Leiedal Intercommunale and Wirral Metropolitan Borough Council. We would also like to acknowledge the assistance provided by our research colleague David McMullan.

REFERENCES

ActiveMaps, 1998, ActiveMaps, http://internetgis.com/.
BORDER, 1998, Business Opportunities for Regional Development and Economic Regeneration, http://www.border.ulst.ac.uk.
Cole, S., 1997, Futures in Global Space, *Futures 1997*, Vol. 29(4-5), pp. 393-418.
Crowder, J., 1996, Mapping the Information Superhighway, in *The AGI source book for GIS,* (London: Taylor & Francis).
ESRI, 1998, MapObjects Literature, http://www.esri.com/library/ whitepagesmo-lit.html.
Heikkila, E. J., 1998, GIS is dead; Long live GIS! *Journal of the American Planning Association,* vol 64(3), pp. 350-360.
MapInfo, 1998, Product Library, http://www.mapinfo.com/*software/library/library.html.*
GI2000, 1998, Geographic Information for Europe http://www2.echo.lu/gi/en/intro/ gihome.html.
VanBrakel, P. A. and Pienaar, M, 1997, Geographic Information Systems: How a World Wide Web presence can improve their availability, *Electronic Library,* Vol. 15(2), pp.109-116.
Waters, R., 1995, Data Sources and Their Availability for Business Users Across Europe, in *GIS for Business and Service Planning,* Longley, P. and Clarke, G. (eds) (Cambridge Geoinformation International).
Webster, G., 1997, Protecting Software, *AGI'97 Conference Proceedings* (AGI & Miller Freeman)

APPENDIX 1 Internet GIS and Mapping Systems

1. IRIS - a knowledge-based system for visual data exploration - http://allanon.gmd.de/and/java/iris/Iris.html.
2. GeoWeb - provides WWW Online Resources for G.I.S., GPS, & Remote Sensing Industries. - http://www.ggrweb.com/.
3. MEL - a distributed data access system (Master Environmental Library) - http://www-mel.nrlmry.navy.mil/homepage.html
4. Interactive Pollution Mapping system is a basic GIS system run by Friends of the Earth - URL http://www.foe.co.uk/cri/html/postcode.html
5. Pima County USA - The Department of Transportation and Flood Control, provide a selection of features at http://www.dot.pima.gov.
6. KINDS - an extensive system, which aims to increase the use of spatial data sets, held by MIDAS http://cs6400.mcc.ac.uk/KINDS/.
7. Safe - a mapping application using the American BC Ministry of Forests FC1 Forest Cover dataset - http://www.safe.com/.
8. Geographic Business Systems (GBS) - Telstra, Australia's national telecom, in association with GBS offers a map-based address locator as part of its on-line directory - http://www.yellowpages.com.au
9. AnchorageMap - by GeoNorth combines mapping information with tax assessment data information. http://www.geonorth.com/products/anchorage/.
10. The MapQuest Java Interactive Atlas interface - uses Java Applet to find and explore cities and towns around the world - http://www.mapquest.com/.
11. ActiveMaps - software is written in Java and allows the presentation of maps and an associated database - http://www.internetgis.com/.
12. TAGIS - The West Virginia Department of Environmental Protection, WVDEP, interactive mapping interface utilises a CGI interface to Arc/Info to generate maps and query attribute data - http://darwin.osmre.gov/test/index.html
13. Irish Environmental Protection Agency's National Freshwater Quality Database - maintains comprehensive, reliable, current data, on the quality of inland waters in Ireland - http://www.compass.ie/epa/tutorial.html

9

Visualising the spatial pattern of Internet address space in the United Kingdom

Narushige Shiode and Martin Dodge

9.1 INTRODUCTION

For the past few years, the growth of the Internet and the services it provides has produced significant effects in many fields including GIS (Couclelis, 1996; Cairncross, 1997). As of July 1998, there are over 36 million host machines that are connected to the Internet (Network Wizards, 1998). Judging from the sales of personal computers and the increase of traffic on the Internet, it is estimated that, by the year 2000, there will be over 400 million users (Gray, 1996; I/PRO, 1998). As popularity of the Internet increases, it is becoming increasingly congested with traffic and interpretation of Internet space becomes crucial (Huberman and Lukose, 1997). Nevertheless, the spatial issues of the Internet have yet to be focused on as an objective of geographical study, despite its increasingly critical role (Batty, 1993; Adams and Warf, 1997).

Based on empirical investigation, this chapter provides an approach to defining and visualising Internet space. It also analyses the spatial distribution of IP addresses registered by different types of organisations. We start by defining Internet space and referring to IP address space. This is followed by details on our data source and the pre-processing steps that we went through. We then explain the results from our data analysis, especially those on different organisation types. We conclude with a summary of the results and discussion on future research.

9.2 BACKGROUND

9.2.1 Internet space and IP addresses

The Internet is often regarded as the collective description of the services provided within this worldwide computer network such as email or the World Wide Web (WWW), but herein we use it simply as a reference to the network itself. Fundamentally, Internet space is not an infinite and unbound electronic domain, but a well defined, closed set, composed of a finite number of elements. It is created by the addressing standard used to identify and locate computers on the network. Each computer is distinguished by a globally unique numeric code called the IP address (short for Internet Protocol address). It is a unique set of 32 bit digit numbers in four blocks of three digits, each taking a value between 0 and 255 (Halabi, 1997).

IP addresses, such as 128.40.59.162, are the fundamental forms of computer addressing used to determine a location within Internet space, enabling computers to communicate with each other. The IP address defines a massive spatial array, containing over four billion unique locations (4,294,967,296 to be precise). In practice, however, the actual usable Internet address space is smaller than the physical maximum because of the way it has been partitioned and allocated (Semeria, 1998).

IP addresses can be compared, in principle, to postcodes that are used in the real world to identify locations for the delivery of letters and parcels. Postcodes define a distinct and widely used form of spatial referencing (Raper *et al.*, 1992). IP addresses are the Internet equivalent, allowing parcels of data to be delivered to the correct computer. IP addresses are usually allocated in large blocks. Taking the postcode analogy further, blocks of IP addresses can be thought of as forming chunks of valuable 'real-estate' on the Internet onto which computers can be 'built'.

9.2.2 GIS and Internet geography

GIS is renowned for its ability to visualise, analyse and model geographical data, applying quantitative geographical methods within a digital environment (Longley and Batty, 1996). GIS relates to the Internet and its geography in two aspects. One is the utilisation of the WWW as a networked GIS delivery media. This is known as Internet-GIS and is becoming increasingly popular with vendors, developers and users (Plewe, 1997).

The other aspect is to define the Internet as a new objective of geographical analysis in conjunction with GIS. Despite the growth of the Internet and its increasing significance, little has been studied on Internet space from this perspective. This would not only encourage a complex structure of Internet space and lack of comprehension towards the Internet domain, but may also hinder the sound development and maintenance of the entire Internet service.

In this chapter, we utilise GIS techniques to better understand Internet space. In particular, we map Internet space onto the real, geographical space by examining the spatial distribution of IP address ownership within the United Kingdom and referring to the location of each point with the corresponding postcode zone. It requires both visualisation capability as well as the analytical functions of GIS to explore the spatial distribution pattern of the 44 million IP addresses. For this purpose, we introduce spatial analysis functions of GIS to investigate the geography of the Internet.

9.2.3 Relevant research

Whether it is a detailed examination of the geography of the Internet or investigation into IP address space from a geographical perspective, there has been little academic research that has actually explored Internet space. There are several interesting studies that cover the Internet geography on a global scale, using data readily available at the national level (Batty and Barr, 1994; ITU, 1997; Press, 1997).

In fact, there are few studies examining the geographical patterns on sub-national scales. A notable exception is the work of Moss and Townsend, Taub Urban Research Center, New York University, who analysed the geography of Internet domain name ownership in the United States (Moss and Townsend, 1997). Similarly, Matrix Information and Directory Services (MIDS, 1998) has mapped Internet hosts and domain names for many countries around the world. MIDS even produced density surface maps of Internet hosts in the United States (MIDS, 1997).

Other visualisation efforts on the Internet are mainly put into mapping cyberspace and network traffic (Dodge, 1998). For instance, Luc Girardin's cyberspace geography visualisation system aims to construct a metaphorical map of the WWW to help people find their way (Girardin, 1996). Similarly, the WEBSOM project creates interactive, multi-scale document maps of Usenet newsgroups (NNRC, 1998). Internet traffic has also been subject to geographic exploration, especially for real-time visualisation of web traffic (Cox and Patterson, 1992; Lamm *et al.*, 1996).

9.3 DATA SOURCE AND PROCESSING

The Internet is in a period of rapid commercialisation and globalisation, undergoing a transition from a network effectively run by the U.S. government and academia to a global communications medium. Consequently, the ownership and management of Internet standards and protocols, such as IP address space and domain names, have recently been subjects of intensive debate (Conrad, 1996; Kahin and Keller, 1997; Shaw, 1997). As the management of Internet space remains controversial, the comprehension of Internet geography is becoming ever more important.

9.3.1 IP space registration

In practical terms, the global IP address space is maintained by the Internet Assigned Number Authority (IANA, 1998) which delegates large blocks of addresses to regional Internet registries (RIRs) to administer (Foster *et al.*, 1997). There are three regional registries responsible for different geographical areas: APNIC (Asia Pacific Network Information Center (APNIC (1998)), ARIN (American Registry for Internet Numbers (ARIN (1998)) and RIPE (Réseaux IP Européens (RIPE (1998)) that covers Europe and surrounding countries. Individual organisations and companies who need an Internet address make requests to the appropriate regional or local registries for the required size range of IP address space. Apart from a small registration fee, IP space is allocated free to the organisations for their exclusive use (Hubbard *et al.*, 1996). However, they have to prove a genuine need for the Internet 'real-estate' as IP space has become a scarce resource (Huston, 1994).

9.3.2 Data source information

As detailed above, RIPE is responsible for the allocation and management of IP address space in Europe and surrounding countries. The RIPE Network Coordination Centre (NCC), based in Amsterdam, maintains a large database of operational Internet information known as the RIPE Network Management Database which contains details on allocation (Magee, 1997). In March 1997, we obtained a copy of the allocation data from the RIPE NCC ftp site with their permission. The data included lists of all the companies and organisations in Europe to which allocations have been made, along with, in most cases, details of two designated contact people in the organisation concerned. A complete postal address was given for the majority of these contact people which we used to geographically locate the IP space allocations. For those with an incomplete address, we obtained the postcode from the street name, if any, or from the name of the organisation.

We viewed these designated organisations and people as the de facto 'landowners' of that block of IP address space and took their postal addresses as the point in geographical space where the IP addresses are located. In order to protect privacy of the individuals and organisations concerned, all personal information was removed, immediately after the designated postal address had been used to give each IP allocation record a geographic co-ordinate.

9.3.3 Limitations of IP address data

There are several limitations with these IP address data. Firstly, the RIPE database is maintained for the purposes of operational management of the Internet and, thus, may not be entirely accurate or suitable for the purposes of geographical analysis. However, it is the only available source of information on the ownership and distribution of IP addresses. Besides, we were quite satisfied with its overall accuracy, including that of the postal address we used to identify the locations.

Secondly, our analysis is based on the spatial distribution of the 'ownership' of IP address space. This may not necessarily be the same as where the space is actually used, i.e. where the Internet linked computers are physically located. In fact, some organisations may have a block of IP addresses registered at the head office and distribute them to different branch offices. There is no direct solution for this problem; however, many of the nation-wide organisations use Intranets to connect their branches and register a relatively small number of IP addresses that are assigned to their Internet gateway machines at their head office or network support division.

Finally, our analysis presents the geographical distribution of IP address ownership at a single point in time. When interpreting our results, it should be noted that, with the rapid growth of the Internet, the IP address space is likely to be subject to constant change. To overcome this problem, we downloaded another set of IP address data from RIPE precisely one year after we obtained the previous data. We intend to gather these data over the coming years to enable time-series analysis.

9.3.4 IP address space in the UK

As of March 1997, the IP address space registry contained 10,660 records of IP space blocks allocated to organisations and companies for use in the United Kingdom. This yielded a total of 44,673,268 unique IP addresses representing about one percent of the total global IP address space. The ownership of these 44.7 million IP addresses were then pin-pointed to a geographical location with the postal address details of their 'landowners'. We matched 99.6% of the IP address space to a geographical location. Interestingly, about one percent of this address space was actually registered to owners located outside the UK, in some twenty-five different countries.

The UK IP address space was categorised into three groups based on the nature of the organisation. These were commercial, government, and academic & non-profit organisations. The breakdown of IP address space by these groups was 77%, 5% and 18%, respectively, as shown in Table 9.1.

Table 9.1 Breakdown of the IP addresses by category

Category	Number of IP Blocks	IP Address Size (mil.)	Percentile in Total Size (%)
Commercial	8176	34228315	76.73
Government	1065	2366768	5.31
Academic and non-profit	1419	8012650	17.96
Total	10660	44607733	100.00

9.4 ANALYSIS OF THE IP ADDRESS DENSITY PATTERNS

The results of our preliminary analysis of spatial patterns of Internet space in the UK are displayed on a geographical framework using continuous density maps. For convenience, we used a subset of the UK data covering only Great Britain, excluding 283 records (674,645 IP addresses) for Northern Ireland, the Channel Islands, the Isle of Man and non-UK registered IP address allocations. In other words, a total of 10,183 allocation records, representing some 43.85 million IP addresses, were used to create density surface maps.

9.4.1 Distribution of IP address blocks in the United Kingdom

The locations of the 10,183 allocation records are mapped in Figure 9.1 with each dot representing one record. There is a wide spatial distribution of dots, covering all parts of Great Britain from Penzance in Cornwall up to the Shetland Islands. As expected, the majority of the allocation records are located in the major cities. Central London clearly stands out, with a very dense cluster of data points, along with the towns in London's hinterland to the west and north. Other notable concentrations include Birmingham, Manchester and Newcastle.

9.4.2 Distribution of IP address space in the United Kingdom

Each of the data points in Figure 9.1 represents widely varying sizes of IP address blocks. To take account of this variation we created continuous density surfaces as an effective means of visualising the data. The number of IP addresses assigned to each point record was used as the data value for creating the density surfaces.

Figure 9.2 shows the IP address density surface for the whole of Great Britain based on 43.85 million addresses. The surface is 'spotty' in appearance, with much of the country effectively flat apart from a relatively few spikes of high IP address density observed in a few urban centres. The highest densities, shown in black in Figure 9.2, are in the range of 9090 to 639,422 IP addresses per square kilometre. There is a concentration of high density IP space spikes evident within and around London. Additional spikes are found in Nottingham and Cambridge with IP address densities of 639,000 and 82,000 per square kilometre respectively. The high IP density in Nottingham is caused by one very large block of address space allocated to an Internet service provider (ISP). Cambridge is noted for a high concentration of computing and research-orientated companies, as well as the university itself. London enjoys a powerful position as the pre-eminent centre for corporate and government headquarters functions.

From Figure 9.2 we conclude that the majority of IP address space is owned by organisations and companies located in the Midland and southern England, particularly in London and its hinterland towns.

9.4.3 Distribution of IP addresses in the Greater London area

Figure 9.3 shows the density surface for the midlands and southern England. County boundaries and motorways are also shown to give context. The 'spotty' density surface is evident. The major peaks of high densities, outside the London region, are in Cambridge, Birmingham and Oxford. Bristol and Portsmouth exhibit rather modest density values.

Figure 9.4 provides a close-up view of Greater London and its hinterland revealing the complexity of IP address density patterns in this region. A particular trend apparent in this map is the much higher densities of IP addresses found on the west side of London compared to those in the east. Central London clearly stands out, with a large oblong-shaped zone of very high Internet address density, peaking at 154,500 per square kilometre. This zone covers the entire City area across to the West-end. There are two other notable high-density areas inside the boundaries of Greater London. The one to the west of central London is the area around Heathrow airport with an IP address density of 18,600 per square kilometre. The other dense patch in north London, with a peak density of 22,900, is due in part to the headquarters of a major ISP.

Looking at the immediate hinterland surrounding London itself, many of its smaller satellite towns have high IP address densities. These satellite towns have experienced considerable growth in computer-related industries in the last ten to fifteen years. Heading north out of London, along the M1 motorway, high densities are apparent in Watford, Hemel Hampstead and particularly St. Albans, with 33,800 IP addresses per square kilometre. St. Albans's high IP address densities are largely caused by the headquarters of the Internet operations for a major telecommunications company. Following the M4 motorway west out of London, several other 'hot-spots' can be observed. These are the towns of Slough, Reading and particularly Bracknell, with 11,900 IP address per square kilometre.

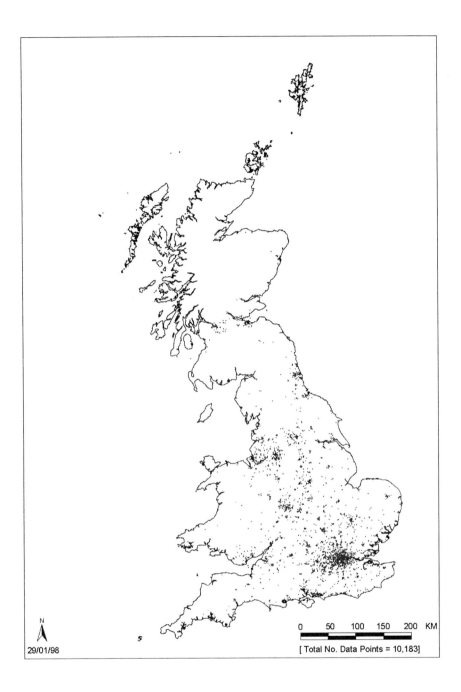

29/01/98

0   50   100   150   200   KM

[Total No. Data Points = 10,183]

Figure 9.1   Distribution pattern of IP address blocks allocated in the United Kingdom.

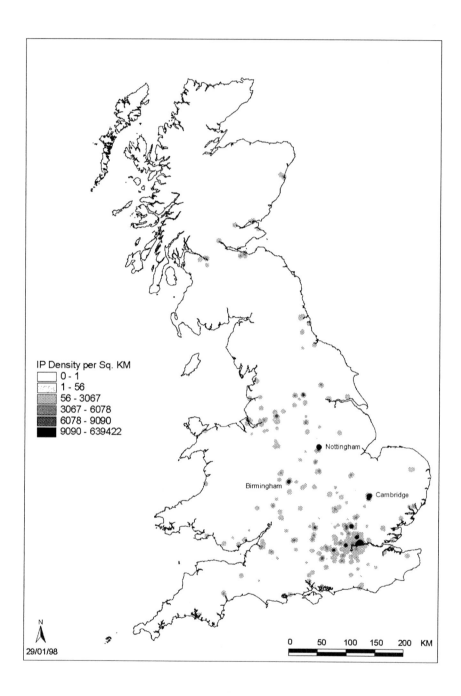

Figure 9.2 IP address density surface covering the United Kingdom.

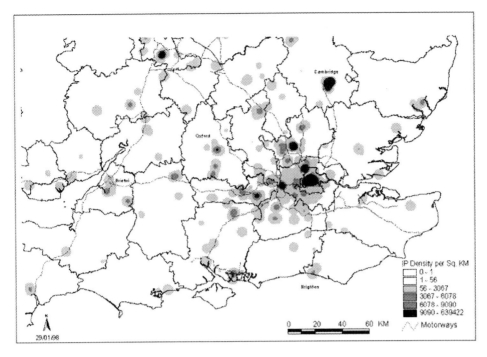

Figure 9.3 IP address density surface of Southern England

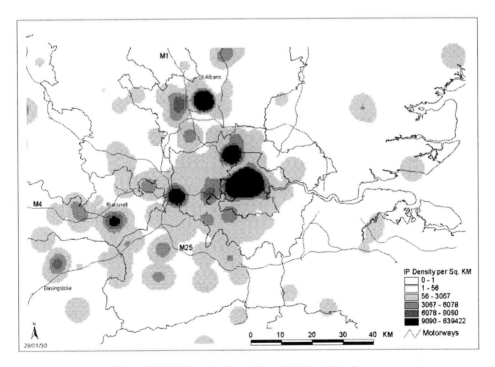

Figure 9.4 IP address density surface of the Greater London area

9.4.4 IP address density maps of different organisations

Figures 9.5-9.7 show IP address surfaces as three-dimensional isometric maps for three different groups; commercial, government, and academic/non-profit organisations. The maps show sharp peaks in this surfaces, with high concentrations in relatively few locations. Also shown is the motorway network to provide some cartographic context. The maps were produced in a three-dimensional visualisation extension to a conventional two-dimensional desktop GIS which allows the user to view the three-dimensional map from any position. As mentioned, the breakdown of these groups was 77%, 5% and 18% respectively. This implies the rapid development and investment by the private sector after the commercialisation of the Internet in the mid-1990s. Although they differ in the share of IP addresses they hold, each group has distinctive role and is by no means negligible. Thus, we mapped them separately onto geo-space and visualised the IP density surface. Interestingly, they exhibit distinctly different patterns.

Figure 9.5 shows the three-dimensional density surface of IP address for commercial organisations. Although we assume the predominant position of London, it actually reveals that one ISP service in Nottingham far exceeds the total IP address density of London area. Figure 9.6 shows the density surface for government organisations. The large peak in the centre shows that Birmingham holds many more IP addresses for governmental use than other densely distributed areas including London and surrounding area, Dundee and Exeter. Finally, Figure 9.7 shows the IP address density surface for academic and non-profit organisations. It displays the dispersed characteristics of academic organisations when compared to Figures 9.5 and 9.6.

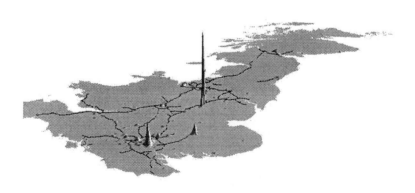

Figure 9.5 Three-dimensional density surface for commercial organisations

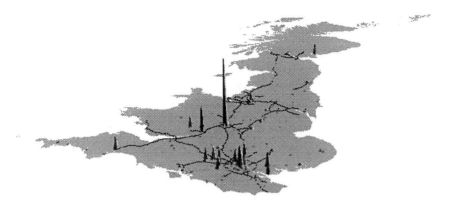

Figure 9.6 Three-dimensional density surface for government organisations

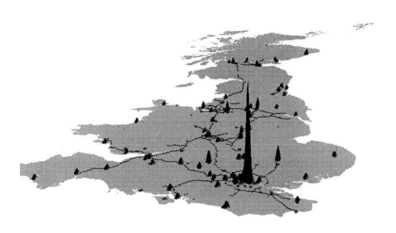

Figure 9.7 Three-dimensional density surface for academic and non-profit organisations

9.5 CONCLUSIONS AND FUTURE PLANS

In this chapter, we have defined the geography of Internet space and analysed it by measuring the density of IP addresses. The majority of the UK's Internet space is owned in a few urban centres. These include central London and the surrounding satellite towns, along with Nottingham, Cambridge and Birmingham. We visualised the variations in ownership of 44 million Internet addresses by creating continuous density surfaces. We also focused on the IP address distribution patterns for different groups of organisations; namely, commercial, governmental, and academic & non-profit.

As the original data from RIPE required a considerable amount of processing, the results still remain at the exploratory stage. However, we are now in a position to undertake further analysis and we believe some interesting results will emerge.

We will develop a GIS model to explain the relationship between the existing socio-economic geography of Britain and the spatial patterns of the Internet 'real-estate' ownership. We are interested in the differences and similarities between the Internet geography and those of the IT services and industries. We want to investigate whether an emerging 'cyber-economy' is concentrating or diffusing the existing patterns of information and technology activities.

We are also trying to perform time-series analysis on the growth in IP address space in the UK. For this purpose, we captured another snap shot of IP address space from the RIPE database on 21 March 1998, exactly one year after the first exercise.

Our final avenue of future research is a comparative study of Internet 'real-estate' examining the spatial patterns in the UK against those in other countries including Japan. We have approached the local Internet registry, Japan Network Information Center (JPNIC) for suitable detailed data on IP address space allocation and are awaiting a response. We are also interested in examining the situation in USA or other European countries, but we may need collaborative partners to achieve the global investigation.

We hope that our study on mapping Internet address space will improve our understanding of the geography of the Internet; making it visible and amenable to spatial analysis. We propose that Internet address space, when related to real-world geospace, could provide a potentially useful surrogate measure of the emerging cyberspace economy. The latest results can be found at our web site, Analysing the Geography of Internet Address Space (Shiode and Dodge, 1998).

ACKNOWLEDGEMENTS

We gratefully acknowledge the RIPE Network Coordination Centre for their permission to use the RIPE database that made this research possible. In particular, we would like to thank John LeRoy Cain at the RIPE NCC for his assistance. However, the views expressed herein are solely those of the authors and in no way reflect those of RIPE or RIPE NCC. Thanks are also due to Sarah Sheppard and Simon Doyle for their help in processing and analysing the data. The digital map and postcode data used in the research was made available by the ESRC and the Midas and UK Borders data services.

REFERENCES

Adams, P.C. and Warf, B., 1997, Introduction: Cyberspace and Geographical Space, *The Geographical Review*, Vol. 87(2), pp. 139-145.

APNIC, 1998, Asia Pacific Network Information Center, http://www.apnic.net.

ARIN, 1998, American Registry for Internet Numbers, http://www.arin.net.

Batty, M., 1993, Editorial: the geography of cyberspace, *Environment and Planning B: Planning and Design*, Vol. 20, pp. 615-616.

Batty, M. and Barr, B., 1994, The electronic frontier: exploring and mapping cyberspace, *Futures*, Vol. 26(7), pp. 699-712.

Cairncross, F., 1997, *The Death of Distance: How the Communications Revolution Will Change Our Lives*, (Boston, MA: Harvard Business School Press).

Conrad, D. R., 1996, Administrative infrastructure for IP address allocation, In the *CIX/ISOC Internet Infrastructure Workshop*, http://www.aldea.com/cix/randy.html.

Couclelis, H., 1996, Editorial: the death of distance, *Environment and Planning B*: *Planning and Design*, Vol. 23, pp. 387-389.

Cox, D. and Patterson, R., 1992, Visualization study of the NSFNET, http://www.ncsa.uiuc.edu/SCMS/DigLib/text/technology/Visualization-Study-NSFNET-Cox.html.

Dodge, M., 1998, An atlas of cyberspace, http://www.cybergeography.org/atlas/

Foster, W. A., Rutkowski, A.M. and Goodman, S. E., 1997, Who governs the Internet?, *Communications of the ACM*, Vol. 40(8), pp. 15-20.

Gray, M., 1996, Internet statistics, http://www.mit.edu/people/mkgray/net/.

Girardin, L., 1996, Mapping the virtual geography of the World Wide Web, in the *Fifth International World Wide Web Conference*, http://heiwww.unige.ch/girardin/cgv/.

Halabi, B., 1997, *Internet Routing Architectures*, (Indianapolis: New Riders Publishing).

Hubbard, K., Kosters, M., Conrad, D., Karrenberg, D. and Postel, J., 1996, Request for comments 2050: Internet registry IP allocation guidelines. ftp://ftp.ripe.net/rfc/rfc2050.txt.

Huberman, B. A. and Lukose, R. M., 1997, Social dilemmas and Internet congestion, *Science*, Vol. 277, pp. 535-537.

Huston, G., 1994, Request for comments 1744: Observations on the management of the Internet address space. ftp://ftp.ripe.net/rfc/rfc1744.txt.

IANA, 1998, Internet Assigned Number Authority, http://www.iana.org.

I/PRO, 1998, CyberAtlas, highlights from around the Web, http://www.cyberatlas.com/.

ITU, 1997, Challenges to the network: tele-communications and the Internet, (Geneva, International Telecommunication Union), http://www.itu.org/publications/index.html.

Kahin, B. and Keller, J., 1997, *Co-ordinating the Internet*, (Cambridge, MA: MIT Press).

Lamm, S. E., Reed, D. A. and Scullin. W. H., 1996, Real-time geographic visualization of World Wide Web traffic, http://www-pablo.cs.uiuc.edu/Publications/Papers/WWW5.ps.gz.

Longley, P. and Batty, M., 1996, *Spatial Analysis: Modelling in a GIS Environment*, (Cambridge: GeoInformation International).

Magee, A. M. R., 1997, RIPE NCC database documentation, ripe-157, *RIPE Network Coordination Centre*, http://www.ripe.net/docs/ripe-157.html.

MIDS, 1997, The Internet by U.S. county, *Matrix Maps Quarterly*, Vol. 402, pp. 40-42.

MIDS, 1998, Matrix Information and Directory Services, http://www.mids.org.

Moss, M. L. and Townsend, A., 1997, Tracking the net: using domain names to measure the growth of the Internet in U.S. cities, *Urban Technology*, Vol. 4(3), pp. 47-59.

Network Wizards, 1998, Internet domain survey, http://www.nw.com/zone/WWW/.

NNRC, 1998, WEBSOM: self-organizing map for Internet exploration, *Neural Networks Research Centre*, http://websom.hut.fi/websom/.

Plewe, B., 1997, *GIS Online*: *Information Retrieval, Mapping, and the Internet*, (Santa Fe, NM: Onword Press).

Press, L., 1997, Tracking the global diffusion of the Internet, *Communications of the ACM*, Vol. 40(11), pp. 11-17.

Raper, J., Rhind, D. and Shepherd, J., 1992, *Postcodes*: *The New Geography*, (London: Longman Scientific and Technical).

RIPE, 1998, Réseaux IP Européens, http://www.ripe.net.

Semeria, C., 1998, Understanding IP addressing: everything you ever wanted to know, http://www.3com.com/nsc/501302.html.

Shaw, R., 1997, Internet domain names: whose domain is this? In *Coordinating the Internet*, Kahin, B. and Keller, J. (eds) (Cambridge, MA: MIT Press).

Shiode, N. and Dodge, M., 1998, Analysing the Geography of Internet Address Space, http://www.geog.ucl.ac.uk./casa/martin/internetspace/.

PART III

Operational Spatial Analysis

10

Putting the Geographical Analysis Machine on the Internet

Stan Openshaw, Ian Turton, James Macgill and John Davy

10.1 INTRODUCTION

Currently most of the proprietary GIS software systems lack sophisticated geographical analysis technology. Attempts over the last decade to persuade the system developers to add spatial analysis functionality has so far failed to have much visible impact. The traditional arguments seemingly still apply; viz. no strong market demand, fear of statistical complexity, lack of suitable GIS-able methods, and a deficiency of statistical skills amongst the GIS system developers.

Various solutions to this dilemma have been suggested and tested; in particular the development of spatial statistical add-ons tied to this or that GIS and the development of standalone statistical packages with either basic GIS functionality or easy linkage to one or more GIS systems. The problem is that these systems mainly serve research needs; see for example, the special issue of *The Statistician,* vol. 47(3), 1998. Unfortunately, most of the potential end-users of geographical analysis methods are not researchers in academia, but involve the far larger numbers of global GIS workers who may often lack the advanced statistical and programming skills needed to apply much of the spatial analysis technology developed for research purposes. Typically the end-users of GIS want easy to use, easy to understand, easy to explain methods with all the complexity hidden away from their gaze. User friendly methods need to be developed so that spatial analysis is no harder to apply than any other part of the GIS tool-kit. Unfortunately, most of the potential users have no good idea of what tools they need because they may not know what are available or what is possible. In addition, there has been considerable misinformation by the GIS vendors as to what spatial analysis is.

There are three further difficulties. First, there is no single dominant GIS and so any spatial analysis technology likely to appeal to end-users has to be what might be termed 'GIS-invariant'. It has to be compatible with all of the world's current GIS systems otherwise the majority of users will never be able to use it. Secondly, most potential end-users may need to be convinced by data trials and tests that a geographical analysis tool is worth using on their data and in the context of their unique operational-institutional-organisational setting. These tests have to be capable of being easily performed as this is

121

probably the principal mechanism for the diffusion of spatial analysis tools and an important device in fostering a spatial analysis culture amongst end-users. If potential users can obtain useful results with their own data at no expense and minimal effort then maybe the case for adding spatial analysis functionality to the GIS tool-kit will be greatly strengthened. Finally, any successful methods have to be available in a platform independent form so they can be moved subsequently on to local user systems.

This chapter focuses on the use of the World Wide Web as a means of providing a platform independent interface to a particular geographical analysis method as a test of viability. The aim is to take a generic and widely applicable point pattern analysis method (the Geographical Analysis Machine), that is currently little used because it is not available in a platform independent form, and convert it into a WWW-based tool so that any user with a web browser could use it. Such a system would be GIS invariant since data from any of the world's GIS systems could be sent to it, analysis performed, the results pre-viewed, and then imported back into any local GIS for further scrutiny. The ideal system would use WWW forms and hypertext to make using it as easy and as automated as possible, whilst providing in depth support and information sufficient to resolve most user questions concerning the application. This strategy would also avoid (or postpone) the traditional problems of porting software onto different hardware for many different operating systems, and the subsequent problems of software support and maintenance. If there is only one geographical analysis machine located on the Internet then most of these problems are avoided. The physical location of the hardware used for this machine could be virtually anywhere in the world, it could be a parallel supercomputer or a workstation. The only remaining problem is how to support an Internet-based dedicated geographical analysis machine and whether the owners of the hardware would always be willing to run a user service on it. The solution could be restriction of the free service to evaluations concerned with testing and prototyping the technology and then allow the program to be downloaded to an interested user's local host. The principal problem resulting from this approach is that the system being downloaded has itself to be portable and platform independent (e.g. Java), or exist in multiple versions for specific target systems (all the species of Unix, NT, windows, OS/2, Mac), or be distributed as source code able to be compiled without difficulty anywhere (i.e. Fortran 77, C, C^{++}).

10.2 A GEOGRAPHICAL ANALYSIS MACHINE

10.2.1 Mark 1 Geographical Analysis Machine

One way of meeting the design objective for end-friendly geographical analysis tools is to develop totally automated geographical analysis methods. The notion of an analysis machine is very attractive. Put your data into a 'geographical analysis machine', click on the 'go' icon, and out come the results. Openshaw *et al.* (1987, 1988, and 1989) described the development of prototype Mark 1 Geographical Analysis Machine (GAM/1) for the analysis of point data for evidence of patterns. This was an early attempt at automated exploratory spatial data analysis that was easy to understand. The GAM sought to answer a very simple practical question; given some point referenced data for a region of interest relating to something of interest to the user *where* might there be evidence of localised clustering if you have no *a priori* idea of where to look due to lack of knowledge about the data and the possible pattern generating processes? This is a generic exploratory geographical analysis task that is widely applicable to many types of data; for example, rare disease data, crime data, burst pipes, customer response, and traffic accidents.

10.2.2 GAM/K Algorithm

The GAM algorithm involves the following steps:

Step 1 Read in X,Y data for population at risk and a variable of interest from a GIS;

Step 2 Identify the rectangle containing the data, identify starting circle radius (r), and degree of circle overlap;

Step 3 Generate a grid covering this rectangular study region so that circles of radius r overlap by the desired amount;

Step 4 For each grid-intersection generate a circle of radius r;

Step 5 Retrieve two counts for this circle (the population at risk and the variable of interest);

Step 6 Apply some 'significance' test procedure;

Step 7 Keep the result if significant;

Step 8 Repeat Steps 5 to 7 until all circles have been processed;

Step 9 Increase circle radius and return to Step 3 else go to Step 10;

Step 10 Create smoothed density surface of excess incidence for the significant circles using a kernel smoothing procedure with an Epanechnikov (1969) kernel and a bandwidth set at the circle radius; this is used to build-up an aggregate surface of significant excess incidence based on all significant circles found in step 7;

Step 11 Map this surface because it is the peaks that suggest where the accumulated evidence of clustering is likely to be greatest.

Note that the original GAM/1 consisted of Steps 1 to 9. Steps 10-11 are the GAM/K version that evolved in the late 1980s (see Openshaw *et al.* (1988, 1989); Openshaw and Craft (1991)). Many subsequent attempts by others to apply GAM have been based on the GAM/1 rather than the more sophisticated GAM/K version. The kernel density surface is very important as a guide as to where to find localised pattern. The accumulation of evidence of many different scales of pattern analysis is a very important but simple mechanism for filtering out false positives and allows the user to focus their attention on the highest peaks. The choice of significance test depends on the rarity of the data, i.e. Poisson, Binomial, Monte Carlo and bootstrap methods can be used. A fuller description is contained in Openshaw *et al.* (1998) and Openshaw (1998).

10.2.3 Reviving GAM/K

An empirical evaluation of eight different clustering methods applied to 50 different rare disease data sets was performed in the early 1990s by the International Agency for Research on Cancer (IARC). The IARC results were sufficiently encouraging to believe that GAM/K was a useful and practical spatial analysis tool; see Alexander and Boyle (1996) and Openshaw (1998) for details of these results.

The code for the original GAM/K still runs on the later day version of the Cray X-MP vector supercomputer (the Cray J90). Initially, efforts were made to port the Cray X-MP code on to a Cray T3D parallel supercomputer with 512 processors. Unfortunately the original Cray code was heavily vectorized and unsuitable for a parallel machine. The algorithm had to be reprogrammed from scratch. Its performance now critically depended on how efficiently it performed the spatial data retrieval. The original GAM/1 used a recursive tree structure (a KD-B tree). In GAM/K for the Cray X-MP this had been replaced by a hash based two dimensional bucket search that was heavily vectorized. Without some form of optimised spatial retrieval the revived GAM/K was estimated to require about 9 days of CPU time on a single J90 processor (or Cray T3D node) to perform a single UK run using census data for 150,000 small areas. A single run could be completed in less than 1 hour on a 512 processor Cray T3D. However, this was clearly unsatisfactory if GAM was ever to become a widely applicable tool.

Fortunately, it was very easy to make GAM/K run much faster. Also it is interesting to note that by 1998 a modern Unix workstation delivers performance levels broadly equivalent to a single Cray J90 processor. Subsequent modifications to the spatial data retrieval algorithm used in GAM/K reduced the 9 days to 714 seconds on a slow workstation. The optimisation strategy was simply to use knowledge of the nature of the spatial search used by GAM/K to minimise the amount of computation being performed. The effects of the code optimisation were dramatic. A test data set of 150,000 points generated a total computation load of 10,498 million megaflops of computation with the naive method but this was reduced to 46 million megaflops by the revised algorithm. GAM/K was now a practical tool from a computational point of view and could be easily linked to any GIS since it no longer needed a supercomputer to run it.

10.3 A WWW IMPLEMENTATION AND EXAMPLE APPLICATION

10.3.1 WWW implementation

The next stage is to consider using simple scripts, HTML forms, and the common gateway interface (CGI) to develop a platform independent user interface to the GAM. The basic design idea was that the user should be able to send their data to GAM, perhaps set a few parameters, and, after a while, the results would be viewable and ultimately sent back to their local site for more detailed study. The WWW would provide a standard GUI with the now common help, buttons, checklists etc. that can be run through any browser.

The first form the user encounters (Figure 10.1) provides a series of boxes for the names of the data files. These files are then uploaded to the Centre for Computational Geography (University of Leeds) web server. The mechanics of this transfer are negotiated between the server and the user's browser.

Next, the user is presented with an HTML form (Figure 10.2) that allows them to specify the parameters of the run. The form uses a combination of type in boxes, pull down menus and check boxes. The browser makes use of the local interface so that the menus look like any other menus typical of a graphical user interface.

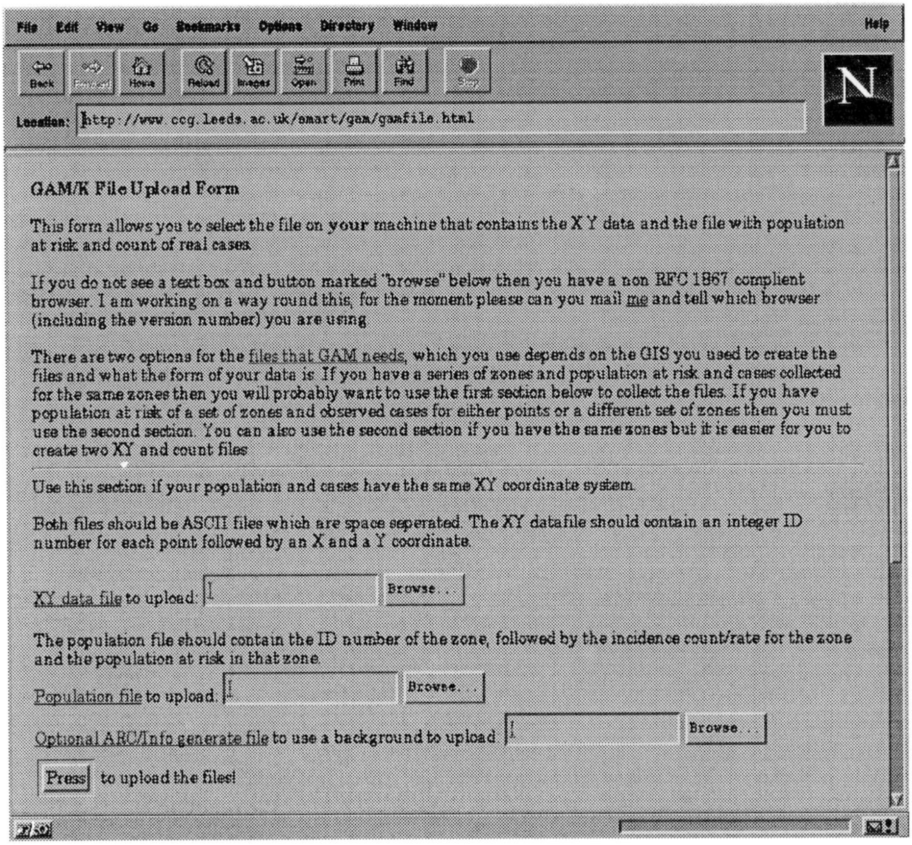

Figure 10.1 The file upload page

The user then proceeds to run GAM on the CCG web server via a CGI script. The final CGI script converts a map of the results into a graphics interchange file (gif) that can be displayed in the user's browser, in exactly the same manner as a normal image.

Other options allow the user to add a simple vector layer (e.g. administrative boundaries, roads etc.) to the output map so that hot spots can be more easily located. The user can then down load the ASCII grid file for import in to Arc/Info or ArcView. Finally, GAM produces a copy of its results in a few alternative standard GIS export formats; for instance as a Arc/Info ASCII grid file. This can then be imported into a local GIS for further study and examination. To test GAM see Openshaw and Turton (1998).

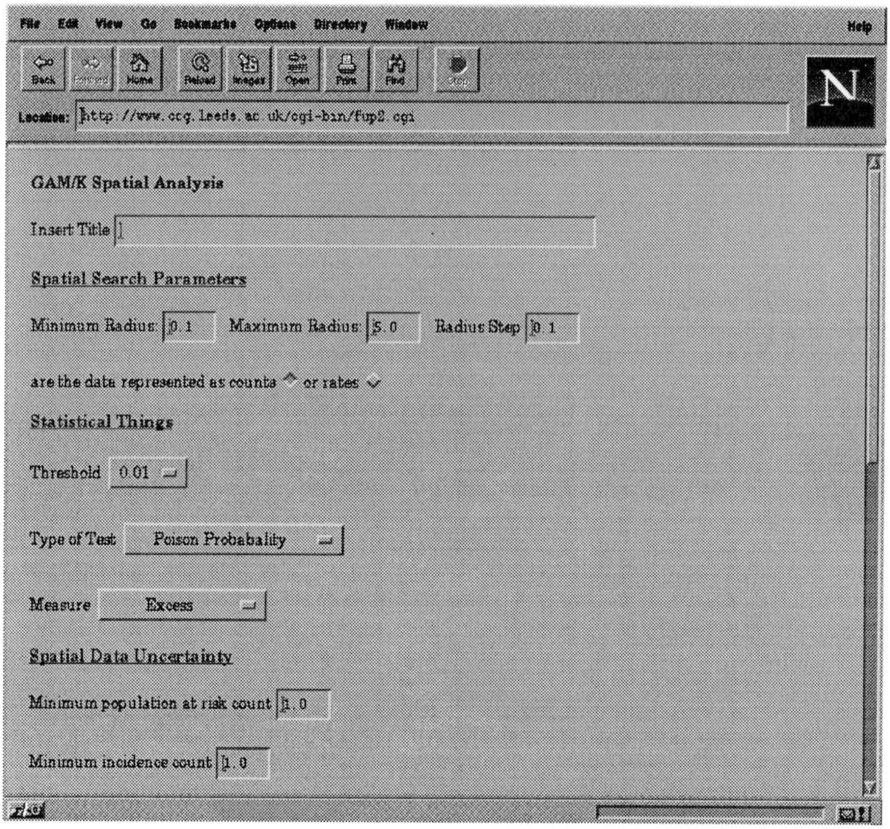

Figure 10.2 Parameters Page

10.3.2 A crime analysis application for a US city

The US Department of Justice Crime mapping and research centre created a crime data set which GAM has been used to analyse. The data was supplied at the US census street block scale of the outlying counties of a US city. The crime data being analysed here are residential burglaries. GAM needs two input files: a file of X,Y co-ordinates and a file of either population at risk or expected values and observed values. The most obvious population at risk for residential crime is the census population although this could be adjusted to reflect socio-economic covariates. These data were sent to WWW-GAM and Figure 10.3 produced. It shows the hot spots detected by GAM in these residential crime data as seen from a web browser. The peaks are where the evidence for unusual localised clustering is strongest. These hot spots are obvious and it is here where further attention should be concentrated. These surfaces can also be converted to the virtual reality modelling language (VRML) and viewed in three-dimensions to allow users who are unfamiliar with choropleth maps to determine the locations of the highest peaks (Figure 10.4). Finally, the results can be exported into ArcView (or any other GIS) for mapping and more detailed investigation.

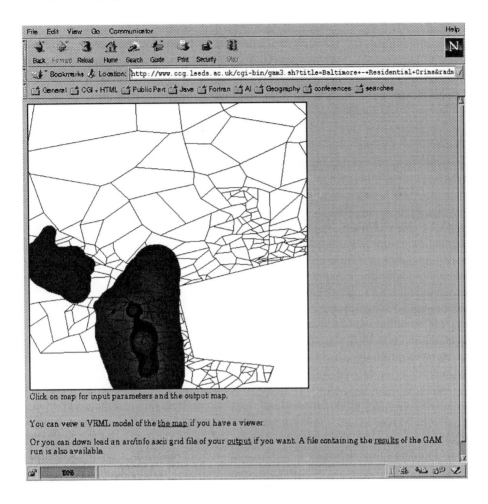

Figure 10.3 Residential burglary hot spots

A residual question is how do you know when a hot spot that appears on the map is not a real hot spot after all? There are four possible answers. One is to feed randomised crime data into GAM and see what sorts of hot spots are found. Typically, they will be weak and not 'highly peaked'; but how high does a peak have to be before it becomes significant? The answer is data dependent and qualitative. It also depends on what the results are to be used for. The safest and simplest strategy is merely to look for the highest peaks. The second answer is to expend more computational time on Monte Carlo simulation to calculate a measure of statistical significance. This will indicate how easy it is to obtain results as extreme as those being observed by running GAM on multiple sets of randomly generated crime data sets with similar incidence to the observed data. The third answer is to keep the results for differing degrees of clustering for training purposes so that the user can be taught how to discriminate between highly significant and unimportant hot spots. The strength of GAM is that it is a visual method of analysis. It is meant to suggest and create new insights in an almost artistic and qualitative kind of way.

The fourth answer is to use GAM purely as an automated procedure and re-install the simple expert system that GAM/K had in 1990 before it was decided to use the human eyeball instead. It is important to remember that there are many different causes of hot spots, many of which are related to the quality of the data being analysed. For instance, no machine-based procedure can, at present, detect 'bad data' for you; you have to do it. One hundred percent automation would so reduce the value of human inputs as to render these spurious causes of clustering totally invisible to the end-user.

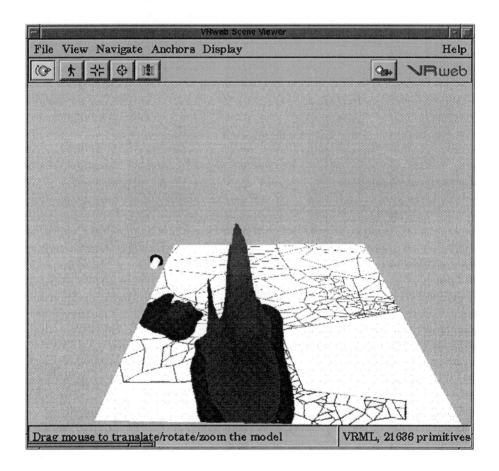

Figure 10.4   Three dimensional VRML view

10.4   FROM FORTRAN 77 TO C⁺⁺ TO JAVA

Logically, the next stage in the development of WWW-GAM is to re-write the program in Java. The reasons for a re-engineering in Java are that: it enables distribution of a completely portable and platform independent version; a modular object-oriented design would facilitate GAM's future deployment in a range of client-server-based spatial analysis scenarios which may well become useful as GAM becomes embedded in more complex and smarter pattern search mechanisms able to explore higher dimensional data spaces; and, there are plans to add animation to GAM and this is most easily done in Java.

The original programming language used for GAM was Fortran 77 and its performance has been substantially improved largely through highly optimised spatial search routines. Despite its great antiquity Fortran code has some advantages. The latest F90 compilers greatly extend the capabilities of Fortran 77 and narrow the gap between Fortran and C. However, if the aim is to squeeze maximum execution time performance out of a code, Fortran 77 code run on a F90 compiler is still best. The simple structure of Fortran allows highly efficient loop optimisation albeit at the expense of programmer productivity. The problem with C and C^{++} is the lack of optimising compilers that can cope with the nature of these codes, whereas Java is even worse because it is currently an interpreted language, although just-in-time (JIT) compilers and other tools narrow the performance gap. However we are entering an era where programmer time is far more expensive than machine time. At present and certainly in the future portability will become the overriding concern.

The translation process proceeded in three stages. First, the Fortran code was translated to a structurally equivalent version in C^{++}, using only a procedural subset of the language. This was a straightforward task which could well have been automated. Secondly, the program was re-engineered in C^{++} to an object-oriented form. The final stage was a straightforward translation from C^{++} classes to Java. Clearly the re-engineering process gains flexibility at the cost (in the short-term) of performance. However, because the re-engineering was fairly coarsely grained it retained the high performance properties of the search algorithms originally developed in Fortran. As a result the loss of performance was likely to be far less than would have been the case if GAM had been object oriented at the circle level. Experiments were carried out to assess this performance impact. Table 10.1 shows the execution times (in seconds) for different sizes of data set. Times include the input of the initial date from files and the output of the final density surface to a file. All results were obtained on a Sun Ultra 30 workstation. Standard Fortran and C^{++} compilers were used, with the highest level of optimisation. The Java version used Sun's JIT compiler.

Table 10.1 Execution times for GAM in various languages on Sun Ultra 30 workstation

Programming Language	9,681 pairs of X,Y	71,911 pairs of X,Y	370,397 pairs of X,Y
Fortran	1.88	10.14	517.6
C^{++}(procedural)	2.61	12.35	493.6
C^{++} (0-0)	2.90	15.13	537.6
Java (JIT)	8.24	27.77	921.5

Given the maturity of Fortran optimising compilers, the performance loss in the other versions is unsurprising, and indeed compares well with what was anticipated. In particular, the performance loss with C^{++} of moving from a monolithic to an object-oriented code structure is irrelevant but this reflects the coarse granularity of the object-oriented code. The performance of Java could be further improved through the use of a native compiler when one becomes available. However, in all cases the execution times are sufficiently small as to suggest that a Java GAM is now a very practicable proposition. The (100% pure) Java version gives the basis for a portable down-loadable version of GAM. Ideally this would be in the form of an applet invoked from within a Java-enabled WWW browser. Unfortunately it is not quite that easy because security restrictions within browsers prevent applets reading or writing the local filestore. For this reason we make GAM available as a Java application in the form of a Jar archive file, which can be downloaded from the GAM web site. After manual extraction of the compiled class files Java GAM can be run using a standard Java interpreter or JIT.

10.5 CONCLUSIONS

This chapter has demonstrated that it is feasible to develop distributed Internet-based approaches to geographical analysis relevant to GIS applications. The same approach is being applied to other geographical analysis tools developed at the University of Leeds with the Geographical Correlates Exploration Machine (GCEM, see Openshaw *et al.* (1990)) expected on stream soon. Other more animation-oriented tools (such as MAPEX, see Openshaw and Perree (1996)) are also planned. The use of the Internet neatly avoids all the problems associated with vendor specific alternatives and allows the development of a genuinely GIS-invariant approach to geographical analysis. However, there are still some unanswered questions. The GAM now exists as a virtual machine but for how long? In the UK the notion of MIDAS style national data servers is well established but the idea of obtaining analysis via a similar route is more novel but presumably it could be added to the existing services. On the other hand, in the not-so-distant future, as hardware costs continue to fall relative to software, the idea of buying a dedicated single function geographical analysis machine (hardware plus software) will not appear as absurd as it once was.

ACKNOWLEDGEMENTS

The research performed here was sponsored by ESRC via grant R237260. The data used here for illustrative purposes was supplied by the US Department of Justice.

REFERENCES

Alexander, F. E. and Boyle, P., 1996, Methods for Investigating localised Clustering of disease (IARC Scientific Publications No 135, Lyon, France).

Epanechnikov, V.A., 1969, Nonparametric estimation of a multidimensional probability density, *Theor. Probab. Appl.,* Vol. 14, pp. 153-158.

Openshaw, S., 1996, Using a geographical analysis machine to detect the presence of spatial clusters and the location of clusters in synthetic data. In *Methods for Investigating Localised Clustering of Disease*, Alexander, F.E., and Boyle, P. (eds), (IARC Scientific Publication No 135, Lyon, France), pp. 68-87

Openshaw, S., 1998, Building Automated Geographical Analysis and Explanation Machines. In *Geocomputation: A Primer*, Longley, P. A., Brooks, S. M., Mcdonnell, R. and Macmillan, B. (eds), (Chichester: Wiley), pp. 95-116.

Openshaw, S. and Craft, A., 1991, Using the Geographical Analysis Machine to search for evidence of clusters and clustering in childhood leukaemia and non-Hodgkin lymphomas in Britain. In *The Geographical Epidemiology of Childhood Leukaemia and Non-Hodgkin Lymphoma in Great Britain 1966-83,* Draper, G. (ed), (London: HMSO), pp. 109-122.

Openshaw, S. and Perrée, T., 1996, User centred intelligent spatial analysis of point data. In *Innovations in GIS 3*, Parker, D. (ed), (London: Taylor and Francis) pp. 119-134.

Openshaw, S. and Turton, I., 1998, Smart spatial analysis, http: //www.ccg.leeds.ac.uk/ smart/intro.htm.

Openshaw, S., Charlton, M., Wymer, C. and Craft, A.W., 1987, A mark I geographical analysis machine for the automated analysis of point data sets, *International Journal of GIS*, Vol. 1, pp.335-358.

Openshaw, S., Charlton, M., Craft, A.W. and Birch, J.M., 1988, Investigation of leukaemia clusters by the use of a geographical analysis machine, *Lancet*, Vol. I, pp. 272-273.

Openshaw, S., Wilkie, D., Binks, K., Wakeford, R., Gerrard, M.H. and Croasdale, M.R., 1989, A method of detecting spatial clustering of disease. In, *Medical Response to Effects of Ionising Radiation*: Crosbie, W.A. and Gittus, J.H. (eds), (London: Elsevier), pp.295-308.

Openshaw, S., Turton, I. and Macgill, J., 1998, Using the Geographical Analysis Machine to analyse long term limiting illness data, *Geographical and Environmental Modelling,* in press.

11

The use of fuzzy classification to improve geodemographic targeting

Zhiqiang Feng and Robin Flowerdew

11.1 FUZZY GEODEMOGRAPHICS

As an application of Geographical Information Systems (GIS) in business, geodemographics has evolved rapidly in the past two decades across Europe and North America. Marketers employ the tools to target consumers on the basis that their residential location can be used to predict their purchasing behaviour. The 'birds of a feather' principle that two people who live in the same neighbourhood are more likely to have similar characteristics than are two people chosen at random underlies geodemographics. This is extended to the principle that people who live in different neighbourhoods with similar characteristics are likely to have similar purchasing characteristics.

In practice, the census unit or postcode is used as the unit of neighbourhood classification. Any basic geographical area can only belong to one cluster. This simplifies the reality that people in the same geodemographic cluster may buy different things, and people in different clusters may share similar purchasing behaviour. Some areal units may be assigned to a cluster without having all that much in common with it, and others may not be assigned to a cluster with which they have much in common. In addition, the boundaries of the areal units do not naturally divide people into totally different groups (Morphet, 1993). With regard to these problems Openshaw (1989a, 1989b) proposed that the fuzziness concept could be integrated into geodemographics in two ways, classification fuzziness and geographical fuzziness. First, a fuzzy approach to geodemographic classification, as described below, may help marketers to target the most relevant consumers whether or not they are all located in one geodemographic cluster. Second, areas which are not in the target clusters, but geographically close to areas in target clusters may also contain potential customers. One way to improve the efficiency of geodemographics is to investigate this fuzziness in geographical space and to identify areas across a boundary from target areas which may share some of their characteristics.

Fuzzy clustering methods have been suggested to cope with the classification problem (Feng and Flowerdew, 1998). The membership values extracted from clustering results indicate to what degree an enumeration district (ED) belongs to various neighbourhood types. An ED in a fuzzy classification can partially belong to different classes at the same time. The EDs which are similar to a target class can readily be identified by looking at their membership values. 'Hardened' classes in which each ED is assigned to that cluster in which it has the highest partial membership, can be used in the same way as conventional hard classes. The membership values can then be used to undertake customer profiling more precisely. Geographical fuzziness can also be handled by adjusting the membership values with reference to spatial effects. The fuzzy clustering method provides a basis for setting up a fuzzy geodemographic system.

Collaboration with market analysts has allowed us to explore the potential of fuzzy clustering in a real geodemographic context. Market survey data collected in Luton provides a basis for comparing the performance of standard geodemographic classification with a fuzzy classification and also to investigate the potential sources of error in geodemographics. In addition to investigating classification and geographical fuzziness effects, we also comment on other issues related to geodemographics, namely the effect of time elapsed since the collection of the data on which the classification is based, and the effects of special-purpose systems.

11.2 FUZZY K-MEANS CLUSTERING

Fuzzy cluster analysis is an extension of traditional cluster analysis. In conventional or 'hard' clustering a sample must be forced into one of several clusters according to some rule or algorithm. However, in fuzzy clustering a sample can partially belong to different groups at the same time by assigning degrees of membership to different groups. Fuzzy k-means clustering (FKM), occasionally called fuzzy c-means, is a generalisation of the conventional k-means algorithm. FKM uses an objective function as the basis of the algorithm. By iteratively minimising the objective function, cluster centroids and membership values are obtained. The objective function is defined as (Bezdek *et al.*, 1984)

$$J(M,\lambda) = \sum_{i=1}^{n}\sum_{j=1}^{k}\mu_{ij}^{m}d^{2}(x_{i},\lambda_{j}) \tag{11.1}$$

The exponent m determines the degree of fuzziness of the solution. When m declines towards its smallest value of 1, the solution is close to a crisp partition. As m increases, the degree of fuzziness will grow till $\mu_{ic}=1/k$ for every sample, when no clusters are found. There is no theoretical or computational evidence distinguishing an optimal m (Bezdek *et al.*, 1984). The optimal value comes from experiments using the real data. Usually, m is chosen in the range 1.5 to 3.0.

By employing FKM to classify the neighbourhood areas an individual area A_i comes out as:

$$(A_{i},\mu_{ij}) \qquad i=1,\dots,N, \quad j=1,\dots,k \tag{11.2}$$

μ_{ij} denotes its membership in cluster type j. If μ_{ij} is 1 then A_i falls precisely in the cluster type j; if μ_{ij} is 0 then A_i is definitely not in cluster type j. Most areas partially belong to several clusters at the same time. The membership functions μ_{ij} observe:

$$\sum_{j=1}^{k} \mu_{ij} = 1 \qquad (11.3)$$

In FKM the membership value of area i in cluster j is defined as:

$$\mu_{ij} = \frac{\dfrac{1}{(d_E(X_i, \lambda_j))^{2/(m-1)}}}{\displaystyle\sum_{j=1}^{k} \dfrac{1}{(d_E(X_i, \lambda_j))^{2/(m-1)}}} \qquad (11.4)$$

where $d(X_i, \lambda_j)$ is the Euclidean distance between the attribute vector x_i and the cluster centroid λ_j.

Mathematically and theoretically, fuzzy clustering has several advantages over conventional clustering methods in that: 1) fuzzy methods can detect the intermediate categories which are common in poorly separated data; 2) fuzzy methods can preserve the information within the classification units. The membership values show how much an individual is compatible with the typical attributes of the clusters; 3) fuzzy methods can achieve better results by identifying more accurate centroids of clusters. By doing this they can reduce the mistakes of misclassification.

Another method used in the research is non-linear mapping (NLM). NLM is a method of transforming data from high dimensional into low dimensional spaces for graphic display. It attempts to minimise the distortion between inter-sample distances. The algorithm was developed by Sammon (1969). For a well separated data set the NLM process can enable the data structure to be clearly visualised. FKM and NLM, used in combination, are powerful tools for multivariate analysis.

11.3 THE LUTON MARKET SURVEY

Luton district in Bedfordshire consists of 16 wards and 340 enumeration districts (EDs). The Luton Market Survey was undertaken by BMRB in September 1996 under the auspices of the Market Research Society's Census Interest Group (CIG) and the Group Market Research department of Whitbread plc. The survey design was based on CACI's ACORN classification system. Among 39 ACORN types in Luton three ACORN types (11, 31 and 44), all of which were well represented in Luton, were chosen corresponding to high, medium and low involvement for a wide range of consumer products. The numbers of EDs in Luton allocated to type 11, 31 and 44 were 23, 25 and 27 respectively.

Table 11.1 Three ACORN types and their labels

ACORN type	Label	Number	Total adults
11	Affluent working couples with mortgages, new homes	23	8263
31	Home owners in older properties, younger workers	25	9085
44	Multi-occupied terraces, multi-ethnic areas	27	9493

Nine wards were randomly selected and two EDs were chosen for each sample ward. This produced six EDs per target ACORN type and eighteen EDs in total. PAF addresses were obtained for sample EDs using AddressPoint and random samples of 50-75 addresses were selected per ED excluding business addresses. The field work was undertaken in-home with randomly chosen adults aged over 15. There was a self-completion questionnaire on product usage and a personal interview on demographics and media use. In total 867 households were validly surveyed with an average of 48 households for each ED.

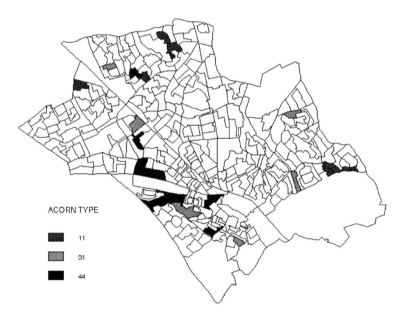

Figure 11.1 The sample EDs and their ACORN types in Luton district

The purposes of the Luton Market Survey were: to understand the changes in census profile in the past five years since it was last recorded; to understand whether differences in census profile really drive differences in product consumption; and to understand if other factors (e.g. relative location) have an effect on an area's behaviour in different consumer markets.

The survey data were firstly analysed to find any relationships between demographic parameters and consumption of products. 17 demographic variables and 13 consumption variables are chosen. Chi-square tests were used to identify whether each demographic variable was significantly associated to consumption of a product. Table 11.2 shows the results: 1 means significant and 0 means not significant at the 95 percent confidence level. Marital status is the most important demographic variable with respect to consumption. The status of chief income earner, age when finishing education, age of respondent and home ownership are also related to the consumption of most products.

Table 11.2 Chi-square test between demographic variables and consumption behaviour

		ice	cof	win	nat	cig	car	hol	tv	vid	dis	ch	wat	sat	sum
C1	age	1	1	1	1	1	1	0	1	1	1	1	*	1	11
C2	work status	1	1	1	1	0	1	1	1	1	1	1	0	0	11
C3	marital status	1	1	1	1	1	1	1	1	1	1	1	0	1	12
C4	children	1	0	0	1	0	1	0	1	1	0	1	0	1	7
C5	respondent status	0	1	1	0	1	1	1	0	0	*	0	*	0	5
C6	respondent's job	0	0	1	0	0	1	1	1	1	1	0	0	1	7
C7	status of chief income earner (CIE)	0	1	1	1	1	1	1	1	*	*	*	*	1	8
C8	position/rank of CIE	0	1	1	0	0	1	0	1	1	1	0	0	0	6
C9	social grade of CIE	0	0	1	0	1	1	1	1	1	1	0	0	1	8
C10	age finished education	1	0	1	1	1	1	0	1	1	0	1	*	1	9
C11	home owned/rented	0	1	1	0	1	1	1	1	1	1	0	1	1	10
C12	home bought outright/mortgage	0	1	1	1	0	1	1	0	1	0	1	0	0	7
C13	home rented/rented free	0	0	0	0	0	1	0	0	0	0	1	0	0	2
C14	length of time in home	0	0	0	1	1	0	1	1	0	1	0	0	0	5
C15	long term illness	0	1	1	0	0	1	1	0	1	1	1	0	1	8
C16	ethnic group	0	0	1	0	1	0	1	1	1	0	0	*	0	5
C17	room number	1	0	1	0	0	1	1	1	1	1	1	0	0	8

(1 significant, 0 not significant at 95 percent confidence level)
ICE: ice cream, COF: coffee, WIN: wine, NAT: natural flavour mineral water, CIG: cigarette, CAR: number of cars, HOL: holiday, TV: number of TV sets, VID: video recorder, DIS: dish washer, CH: central heating, WAT: TV watching, SAT: satellite TV watching. * means chi-square unavailable

11.4 SOME INITIAL RESULTS

11.4.1 Matching between ACORN types and consumption clusters

A basic assumption underlying geodemographics is that the social identity derived from demographic characteristics is related to customer behaviour and thus can be used to predict consumption patterns. Quantitatively, consumption patterns can be measured to allow investigation of association with geodemographic types. Therefore we classified the 18 EDs into three clusters according to the consumption data (Figure 11.2, Table 11.3), in order to compare them with the three ACORN types. The hypothesis is that ACORN types match consumption clusters. The 13 product consumption variables were ice cream, coffee, wine, natural flavour mineral water, cigarettes, cars, holidays, TV sets, videos, dishwashers, central heating, TV watching, and satellite TV watching. Z-scores were obtained for each variable to avoid the bias which may arise in clustering from different variances of variables. The FKM method was used to cluster the data. The centre values of each group clearly show its consumption patterns. For example, cluster 3 has high penetration in all products, but cigarettes, cluster 1 low penetration in all products, but medium in ice cream, cigarettes, number of cars, central heating and satellite TV watching. The results are very similar to the ACORN clustering of geodemographic data. With reference to ACORN types, cluster 1 is related to type 44, cluster 2 to 31, and cluster 3 to 11. However there are three EDs which do not fit into this linkage correctly. ED 5 is in ACORN type 31 and consumption cluster 3, ED 7 in ACORN type 44 and consumption cluster 2, and ED 16 in ACORN type 31 and consumption cluster 1.

The survey data was also processed using NLM methods to give a visual picture of the consumption clusters and to compare them with ACORN. The chart (Figure 11.2) shows that the three clusters are not totally distinct but that EDs similar to each other are assigned to the same cluster. This is especially true for consumption clusters 1 and 3. Cluster 2 shows some transitional features and may share some attributes with cluster 1 or cluster 3. For example ED 7 is similar to cluster 1 and ED 13 is similar to cluster 3.

Table 11.3 Results of FKM analysis for Luton consumption survey data

ED	cluster 1	cluster 2	cluster 3	Hardened cluster	ACORN
1	0.024	0.101	0.876	3	11
2	0.038	0.149	0.813	3	11
3	0.101	0.224	0.675	3	11
4	0.575	0.366	0.059	1	44
5	0.036	0.143	0.821	3	31
6	0.018	0.074	0.908	3	11
7	0.22	0.666	0.114	2	44
8	0.597	0.325	0.077	1	44
9	0.135	0.705	0.16	2	31
10	0.765	0.192	0.043	1	44
11	0.019	0.065	0.917	3	11
12	0.201	0.67	0.129	2	31
13	0.157	0.485	0.358	2	31
14	0.047	0.202	0.751	3	11
15	0.54	0.284	0.176	1	44
16	0.769	0.199	0.031	1	31
17	0.074	0.74	0.186	2	31
18	0.886	0.089	0.025	1	44

The membership values derived from FKM analysis give us quantitative measures of how similar these EDs are to each other on the basis of all three clusters. ED 7 has a membership value of 0.67 in cluster 2 which makes us assign it to that cluster. In addition, it has a membership value of 0.22 in cluster 1. ED 13 has 0.36 membership value in cluster 1 second to 0.49 in cluster 2. This shows that FKM and NLM work very well with the customer data and they confirm each other's results.

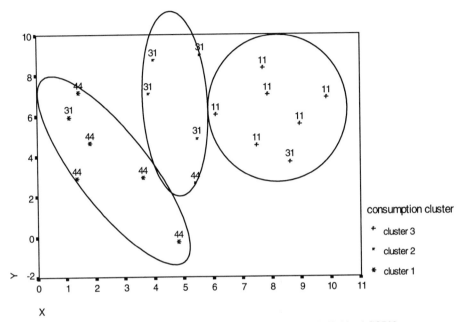

Figure 11.2 Scatter plot for NLM results from Luton survey data labelled by ACORN types

11.4.2 Matching fuzzy geodemographic clusters and consumption clusters

Given that the aim of this chapter, as discussed in section 11.1, is to assess the benefits of using fuzzy geodemographic clustering, the next step is to cluster the 18 EDs on the basis of census data using FKM. We intend to see if the results of FKM improve on standard classification methods and are better related to consumption clusters. Below we use the term group to refer to FKM clusters based on census data to avoid confusion. We chose 44 variables covering mainly economic, demographic, and social characteristics (Openshaw & Wymer, 1995) which were then transformed to obtain Z-scores which were used in the FKM analysis. These variables are similar, but probably not identical, to those used in ACORN; the intention was to produce a fuzzy classification comparable to ACORN, but the exact methodology used in the latter system is not available. The number of groups was set to three for comparison with ACORN. It turns out that these groups match well with consumption clusters: group A to cluster 1, group B to cluster 2 and group C to cluster 3. However, there are two exceptions, ED 7 and ED 16. ED 7 belongs to group A and cluster 2 and ED 16 to group B and cluster 1. On the other hand, the fuzzy classification performs better than ACORN with regard to ED 5, assigning it to the group which matches its consumption behaviour.

By and large, ACORN types still fit the consumption clusters well five years after the census. The three ED exceptions could be for one of the following reasons: 1) time effect: there may have been significant changes since 1991 in these EDs; 2) classification fuzziness: conventional clustering ignores the fuzziness in attribute space and tends to misclassify some units to the wrong groups; 3) other factors influencing customer behaviour are not taken into account when clustering census data. One of these factors is the spatial effect which is worth further exploration. The fuzzy clustering methods seem to improve results a bit by assigning ED 5 to the correct consumption cluster. Two EDs, 7 and 16, remain unexplained.

Table 11.4 Membership values of EDs produced by FKM analysis of Luton census data

ED	Group A	Group B	Group C	Hardened Group	ACORN
1	0.058	0.204	0.738	C	11
2	0.038	0.121	0.841	C	11
3	0.119	0.304	0.576	C	11
4	0.748	0.2	0.052	A	44
5	0.05	0.257	0.693	C	31
6	0.039	0.133	0.827	C	11
7	0.645	0.255	0.1	A	44
8	0.577	0.316	0.107	A	44
9	0.137	0.675	0.188	B	31
10	0.778	0.148	0.074	A	44
11	0.016	0.048	0.936	C	11
12	0.133	0.722	0.145	B	31
13	0.066	0.838	0.096	B	31
14	0.059	0.168	0.772	C	11
15	0.55	0.255	0.196	A	44
16	0.136	0.754	0.11	B	31
17	0.062	0.769	0.169	B	31
18	0.578	0.286	0.135	A	44

11.4.3 The time effect

There was a question in the Luton Survey questionnaire which was related to length of time at current home (LTH). The five categories of answer are: less than one year, 1-4 years, 5-9 years, 10-19 years and 20 years or more. Thus the respondents in the first two categories are in-migrants since 1991. This variable can be used to explore the time effect, e.g. whether the failure of EDs 7 and 16 to fit the predictions of geodemographics is because the two EDs have changed their population composition since the 1991 census.

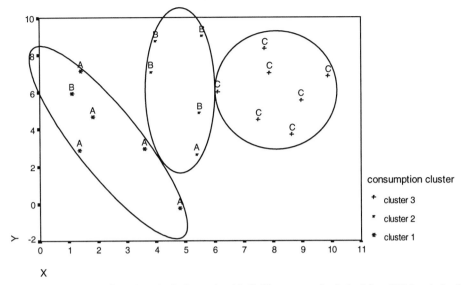

Figure 11.3 Scatter plot of NLM results for Luton data labelled by group codes derived from FKM analysis of census data

In fact, ED 14 has the highest proportion of in-migrants with 20 and 30 percent of residents respectively living at their present home for less than 1 year and 1-4 years. That means that half the residents in ED 14 are newcomers since 1991. In contrast, ED 7 and ED 16 have fairly low proportions of in-migrants since 1991. For ED 7 less-than-one-year residents account for 10 percent and 1-4 year residents for 23 percent, summing to 33 percent. For ED 16 less-than-one-year residents account for 9 percent of respondents and 1-4 year residents for 17 percent summing to 26 percent. The figures are lower than the average 35 percent.

These percentages may not explain the time effect correctly because it is not only the quantities of in-migrants, but differences between the socio-economic characteristics of in-migrants and existing residents that would alter the demographic structure of the neighbourhoods.

Chi-square tests between the attribute LTH and other demographic variables were conducted separately for each ED. The LTH variable is merged into three categories, 1: less than 5 years, 2: more than 5 years and 3: no answer. There are 15 EDs in which there is a significant relationship between LTH and the variable home owned/rented. For ED 14 which has the largest number of in-migrants LTH does not show significant association with many other demographic variables. Actually only four demographic variables are associated with LTH and these four do not include the two most important variables, marital status and status of chief income earner, which are associated with product consumption. This indicates that for ED 14 in-migrants possess generally similar characteristics to previous residents.

In contrast, the LTH for ED 16 is significantly associated with 7 other demographic variables, which is the largest association number among all 18 EDs. The more important thing is that these demographic variables are those which have significant associations with consumption behaviour, marital status, age, status of chief income earner, age finished full-time education and home owned/rented. Therefore it can be assumed that there is a considerable time effect for ED 16 and that the new residents have changed the socio-economic characteristics of the area. ED 7 only has 4 significant variables, age, status of respondent, status of chief income earner and home owned/rented, associated with the in-migrant variable. Therefore the same conclusion for ED 16 cannot be applied to ED 7.

Table 11.5 Association between length in present home and other demographic variables

Area	c1	c2	c3	c4	c5	c6	c7	c8	c9	c10	c11	c12	c13	c14	c16	c17	sum
7	0	0	0	0	0	0	0	0	0	0	1	0	0	0	0	0	1
16	1	0	1	0	0	1	0	0	1	1	1	0	0	0	0	0	6

(1: significant, 0: not significant at 95 percent confidence level)
Fisher's exact test was used when expected values were too low for chi-square statistics to be reliable

11.4.4 The spatial effect

The next stage of the research is concerned with analysis of the spatial effect. We can identify this effect if the variation in consumption patterns between EDs is related not only to their own demographic characteristics, but also to the characteristics of their surrounding EDs. The spatial effect would suggest that if the surrounding EDs have similar characteristics to the target ED the consumption pattern will not change. If the surrounding EDs are in classification groups with higher penetrations for a certain product these will have a positive influence on the target ED. If the surrounding EDs are in groups less likely to buy the product, the influence is negative.

The spatial effect is modelled by adjusting membership values of EDs. The adjusted membership value in a cluster j is defined as:

$$\mu'_i = \alpha\mu_i + \beta\frac{1}{A}\sum_{j=1}^{n} w_{ij}\mu_j \qquad (11.5)$$

where μ' and μ are respectively the new and old degrees of membership of an ED in a cluster, and n is the number of EDs contiguous to EDi. α, β are respectively weights for unadjusted cluster membership and the mean of membership values of surrounding EDs, which sum to 1. The weight could be based on distance between ED centroids or the length of common boundary between EDs, or both. Following Cliff and Ord (1973), we define the weight:

$$w_{ij} = d_{ij}^{-a} p_{ij}^{b} \qquad (11.6)$$

where d_{ij} refers to distance between the centroids of ED i and the jth contiguous ED, p_{ij} is the length of common boundary between them, a, b are parameters. Positive values of a, b give greater weights to pairs of EDs which have shorter distances between their centres and which have long common boundaries. In addition, w_{ij} must be scaled so that the average of weighted membership values is in the range of 0 to 1; this is the role of parameter A.

The fuzzy discriminant method is used to compute the membership values of all EDs in Luton district. Equation 11.4 is applied to the data with the distance defined as that between national centroids of principal components of ACORN types as centre values and the principal component scores of each ED. The results perfectly coincide with the ACORN types with hardened groups matching the types. New membership values of sample EDs are then computed. These new membership values are not significantly different from the original figures. In this case therefore spatial effects do not have much influence on the sample EDs.

11.4.5 Weighted fuzzy classification

The information that demographic variables are associated with consumption of products in Table 11.2 can be used to modify the clustering. Because the principal components provided by CACI do not have demographic attributes attached they cannot be used. Instead we use the 44 variables mentioned above to construct a fuzzy classification, weighting them in accordance with their relationship to the consumption variables as measured by the chi-square results. When this weighted fuzzy classification is produced, ED 7 appears as intermediate between Group A, which is equivalent to the ACORN type 44 to which it belongs, and Group B, which fits with the consumption profile for the ED. So the weighted fuzzy classification approaches a correct allocation for this ED too.

Table 11.6 Membership values from FKM analysis on the weighted census data for problem EDs

ED	Group A	Group B	Group C	Hardened group	ACORN
5	.222	.249	.529	C	31
7	.467	.409	.124	AB	44
16	.205	.712	.082	B	31

11.5 CONCLUSION

Fuzzy geodemographics was first proposed almost ten years ago but has not been fully tested and explored. Our work is a further step toward the setting up of fuzzy geodemographic systems. From our work on the Luton Survey data some conclusions can be drawn. Firstly the ACORN classification system performs reasonably well with reference to customers' consumption patterns. This suggests that the demographic characteristics of an area can be used fairly effectively to predict differences between consumption behaviour. Next, although only slightly, fuzzy methods do improve the classification results and have a better association with consumption patterns. However, the results of our research do not prove the superiority of the fuzzy method to the conventional ACORN classification because the classification is data dependent. The ACORN system is based on national data, while our classification applies only to local data. In addition, differences in the pre-processing of the raw data plays a role in generating different results. Thirdly, the time effect is an important factor because a big proportion of in-migrants may greatly change the structure of residents in a small area like an ED. Between 1991 and 1996 at least one fifth of residents were new movers to the sample areas. However, most EDs in Luton district did not show major changes in population characteristics. An exception is ED 16 in which new in-migrants are considerably different from previous residents. The time effect may not be as great as people thought. The time effect is geographically dependent, e.g. different areas being influenced differently. Fourthly, the spatial effect was not a big issue in Luton data since it did not change the basic classification pattern. Finally, data from the survey can be used to weight the raw data and improve the fuzzy clustering. The last point is very useful. The basic neighbourhood classification is general purpose. In real applications external customer information can be used to modify the clusters and therefore to tailor the classification to fit the product to be marketed. Since nowadays customer information is not a big problem, a small sample of customers can be employed to improve the general purpose classification.

ACKNOWLEDGEMENTS

The authors are grateful to the Market Research Society Census Interest Group Working party on Validation of Geographical Data and to Whitbread Group Market Research, co-sponsors of the Luton survey, and to Barry Leventhal of Berry Consulting and John Rae of CACI for the supply of the Luton Market Survey data and ACORN classification data. BMRB International conducted the fieldwork. We also acknowledge the use of Crown Copyright 1991 Census Small Area Statistics and digitised boundary data purchased by ESRC and JISC for the academic community, and accessed through the Midas system at Manchester Computer Centre.

REFERENCES

Bezdek, C. J., Ehrlich, R. and Full, W., 1984, FCM: The fuzzy c-means clustering algorithm. *Computer and Geosciences,* Vol. 10, pp. 191-203.
Cliff, A. D. and Ord, J. K., 1973, *Spatial Autocorrelation.* (London: Pion)
Feng Z. and Flowerdew R., 1998, Fuzzy geodemographics: a contribution from fuzzy clustering methods *Innovations in GIS 5,* Carver, S. (ed) (London: Taylor and Francis).
Morphet, C. 1993, The mapping of small-area census data - a consideration of the effects of enumeration district boundaries. *Environment and Planning* A, Vol. 25, pp. 1267-1277.
Openshaw, S., 1989a, Making geodemographics more sophisticated, *Journal of the Market Research Society*, Vol. 31, pp. 111-131.
Openshaw, S., 1989b, Learning to live with errors in spatial databases, *The Accuracy of Spatial Databases,* M. Goodchild and S. Gopal (eds) (London: Taylor and Francis).
Openshaw, S. and Wymer, C., 1995, Classifying and regionalizing census data. *Census User's Handbook,* Openshaw, S. (ed) (Cambridge: Geo Information International).
Sammon, J. W., 1969, A non-linear mapping for data structure analysis. *IEEE Transactions for Computers* C18, pp. 401-409.

12

Application of pattern recognition to concept discovery in geography

Ian Turton

12.1 INTRODUCTION

Openshaw (1994) argues that as the amount of data that is collected as a result of the GIS revolution increases, geographers must start to apply new methods to these new data riches. It is no longer enough to merely catalogue the data and draw simple maps of it. It is also no longer acceptable to use crude statistical measures that average over a whole map or region and in so doing throw away the geographical content of the data. In other words, geographers must generalise or drown in the flood of spatial data that has increased many fold during the 1980s and 1990s and which will continue to grow into the next century. As the amount of data grows, it becomes increasingly difficult for humans to find the time to study and interpret the data; the only solution is to pass more of the routine analysis to computers leaving the researcher with more time to study the truly interesting parts of the output.

This chapter is a first attempt to apply these ideas to a geographical data set. One data set will be studied in detail though the ideas and methods developed will be applicable to many other data sets. The data set selected for this study is a population density surface derived from the 1991 census of population by Bracken and Martin (1989). Using this data set the aim is to take the data-poor geographical theories of urban social structure of the first half of the century and make use of the data-rich environments of the 1990s to test the theories in a general and robust manner. To achieve this pattern matching techniques used in computer vision and other fields will be applied to raster data of population density and socio-economic variables for Great Britain.

12.2 REVIEW OF URBAN SOCIAL STRUCTURE

Over time as a city grows different areas become associated with different types of population and this leads to systematic relationships between geographic and social space. Bourne (1971) says, "All cities display a degree of internal organisation. In terms of urban space, this order is frequently described by regularities in land use patterns." These observations lead to questions about how these patterns can be modelled to allow comparisons to be made between cities and to attempt to give insights into the growth and formation of these patterns. Much of the defining work on these questions was undertaken by the Chicago ecologists who were concerned with the differences in environment and behaviour in different parts of the city. From this descriptive research into behaviour in various parts of the city grew an interest in the general structure of the city and its evolution over time. The majority of this work was confined to the study of the large industrial city of Chicago. The work of the Chicago group can be traced back to the work of Hurd (1903) who developed several theories of urban expansion which stressed two main methods of growth: central and axial growth. The zonal model developed by Burgess (1925) is based solely on the central growth element and radial expansion. This is closely linked to the general assumptions of impersonal competition of ecological theory. The zonal model came about almost as an aside in the discussion of how urban areas expand. The sectoral model developed by Hoyt (1939) has a much narrower focus than the zonal model of Burgess, being primarily concerned with the distribution of rental classes. Based upon a block by block analysis of changes in a variety of housing characteristics in 142 US cities, Hoyt concludes that: "The high rent neighbourhoods do not skip about at random in the process of movement, they follow a definite path in one or more sectors of the city. ... the different types of residential areas tend to grow outward along rather distinct radii, and a new growth on the arc of a given sector tends to take on the character of the initial growth of that sector." Again the spatial expression of the model is an aside to the main thrust of the model which is the dynamics of rental patterns.

12.3 IMAGE ANALYSIS AND COMPUTER VISION METHODS

There are many possible definitions of image processing and computer vision. One of the most common is that image processing refers to processing an image (usually by a computer) to produce another image, whereas computer vision takes an image and processes it into some sort of generalised information about the image, such as labelled regions.

The aim of using image processing and computer vision within the context of this work is twofold. First the computer program must take a raster population density map of Great Britain, process it to provide a clean image for later processes to work with, then segment the image to find the urban areas. A second stage must extract these areas and use the locations of the towns and cities discovered in the first stage for classification against existing theoretical models of urban structure and also as inputs to a classifier that attempts to discover new structures within the British urban environment.

X gradient			Y Gradient		
-1	0	1	-1	-2	-1
-2	0	2	0	0	0
-1	0	1	1	2	1

Figure 12.1 Sobel filters

A technique that can be applied to the detection of urban areas with a population density surface is the determination of the gradient of the surface. This is analogous to the differentiation of a mathematical function. To determine where the maxima of a function are, the function can be differentiated. Where dy/dx is zero the function is either at a maximum, minimum or a point of inflection. To distinguish between these the function is differentiated again to give the second differential (d^2y/d^2x) which is negative at maxima. A similar process can be applied to images. The usual way to calculate a gradient of an image is to apply a pair of Sobel filters (see Figure 12.1) to the image. Figure 12.2 shows an example of how this pair of filters are convolved with an image to give an X and Y gradient, which are then combined by using the sum of the absolutes to give the gradient of the image.

In the upper left hand section of Figure 12.2 the input image which contains a peak in the centre of the image can be seen. In the upper right hand corner we see the Y or vertical gradient of the input image. The lower left hand section contains the X or horizontal gradient of the image. In the lower right hand corner is the sum of the absolutes of the two gradients. Both the horizontal and vertical gradients show a zero crossing at the centre, however this information is lost in the sum of the absolutes of the gradients. To discover if the zero crossings found in the gradients are maxima or minima it is necessary to calculate the second derivative of the image. This is done by applying the same Sobel templates to the X and Y gradients. The two second derivatives can be combined by summing the absolutes of the two gradients. If the sign of the output is required then the sign of the sum can be set to the sign of the largest absolute value of the two gradients.

```
1   1   1   1   1   1   1   1         -    -    -    -    -    -    -   -
1   2   3   4   4   3   2   1         -    4    8   11   11    8    4   -
1   2   3   4   4   3   2   1         -    3    4    4    4    4    3   -
1   3   4   5   5   4   3   1         -    3    5    7    7    5    3   -
1   3   4   6   6   4   3   1         -    0    1    3    3    1    0   -
1   3   4   6   6   4   3   1         -    0   -1   -3   -3   -1    0   -
1   3   4   5   5   4   3   1         -   -3   -5   -7   -7   -5   -3   -
1   2   3   4   4   3   2   1         -   -3   -4   -4   -4   -4   -3   -
1   2   3   4   4   3   2   1         -   -4   -8  -11  -11   -8   -4   -
1   1   1   1   1   1   1   1         -    -    -    -    -    -    -   -


-    -    -    -    -    -    -   -         -    -    -    -    -    -    -   -
-    6    6    3   -3   -6   -6   -         -   10   14   14   14   14   10   -
-    9    8    4   -4   -8   -9   -         -   12   12    8    8   12   12   -
-   11    9    5   -5   -9  -11   -         -   14   14   12   12   14   14   -
-   12   11    7   -7  -11  -12   -         -   12   12   10   10   12   12   -
-   12   11    7   -7  -11  -12   -         -   12   12   10   10   12   12   -
-   11    9    5   -5   -9  -11   -         -   14   14   12   12   14   14   -
-    9    8    4   -4   -8   -9   -         -   12   12    8    8   12   12   -
-    6    6    3   -3   -6   -6   -         -   10   14   14   14   14   10   -
-    -    -    -    -    -    -   -         -    -    -    -    -    -    -   -
```

Figure 12.2 First derivative of an image using a Sobel filter

By combining the information obtained from the two steps in the gradient process peaks can be detected. Areas that have a low gradient and a high negative second derivative are easily selected after processing; thresholding can be applied to limit the areas selected. This makes use of the fact that, following aggregation, the population surface is relatively smooth and that urban areas are maxima of the surface. This means that no assumptions about the maximum of population density of the urban area are required and that both large cities and smaller towns can be detected by the same process. As can be seen in Figure 12.3, a range of city sizes from London to Aberdeen are detected. This figure was constructed by calculating the second derivative of the population density surface as described in above and then applying a threshold that removed all cells with a value greater than -45. This figure was obtained by inspection of a histogram of the second derivatives and was a clear break, see Turton (1997) for a fuller description.

Bounding boxes for each area were then calculated so that census data could be extracted at a 200 metre resolution for the urban areas. Any box with an area of less than one kilometre square was eliminated from the study since it was felt that these areas would be too small to show any internal structure even at the 200 metre resolution. The next step was to determine which town was within each box. A gazetteer sorted by population size was used in conjunction with a point in polygon program to determine the largest town in each box. The output was then modified by hand where two large urban areas had been conglomerated by the extraction program, for example Leeds and Bradford are a single output box, but only Leeds was inserted by the point in polygon method, so Bradford was inserted by hand. Table 12.1 gives a list of the urban areas shown in Figure 12.3.

Figure 12.3 Second derivative of GB population, thresholded at -45

Table 12.1 List of towns extracted from population surface

Aberdeen	Dundee	Edinburgh	Greenock
Glasgow	Airdrie	Coatbridge	Motherwell
East Kilbride	Newcastle	Sunderland	Gateshead
Carlisle	Hartlepool	Stockton-on-Tees	Middlesborough
Darlington	Scarbourgh	Barrow in Furness	Harrogate
York	Blackpool	Skipton	Leeds-Bradford
Burnley	Hull	Preston	Blackburn
Accrington	Halifax	Dewsbury	Wakefield
Southport	Huddersfield	Rochdale	Bolton
Bury	Scunthorpe	Manchester	Grimsby
Barnsley	Wigan	Doncaster	Liverpool
Leigh	St Helens	Birkenhead	Sheffield
Rotherham	Warrington	Widnes	Runcorn
Ellesmere Port	Macclesfield	Chesterfield	Chester
Mansfield	Crewe	Stoke on Trent	Nottingham
Derby	Stafford	Norwich	Leicester
Lichfield	Birmingham-Wolverhampton	Peterborough	Lowestoft
Nuneaton	Corby	Coventry	Kidderminster
Redditch	Leamington Spa	Northampton	Cambridge
Worcester	Bedford	Ipswich	Milton Keynes
Luton	Stevenage	Colchester	Cheltenham
Gloucester	Aylesbury	Harlow	Hemelhempsted
St Albans	Chelmsford	Oxford	London
Swansea	High Wycombe	Newport	Basildon
Southend	Swindon	Slough	Cardiff
Maidenhead	Bristol	Reading	Gravesend
Barry	Bracknell	Chatham-Gillingham	Ramsgate
Bath	Farnborough	Woking	Maidstone
Basingstoke	Aldershot	Guildford	Tunbridge Wells
Crawley	Southampton	Hastings	Petersfield
Brighton	Portsmouth	Worthing	Eastbourne
Bournemouth	Exeter	Torbay	Plymouth
Weston-Super-Mare			

Once features have been extracted the process moves on to the detection of patterns within the features and the classification of the features discovered. At this stage it is important that possible matches between an observed urban area and a theoretical template or another observed area are not overlooked due to changes in orientation, size or possible distortions caused by local topology. It is therefore necessary that the second stage of the process makes corrections for these differences without losing sight of the underlying structure of the area.

To achieve this aim it is possible to make use of many techniques that are based on template matching. These techniques take a pattern template that is being sought in the image and compare it to the image making use of a metric such as the sum of the absolute errors or the sum of the square of the errors between the template and the image. When an image closely resembles the template then these measures will be small whereas if the template is very different from the image then the error measures will be larger.

There are many problems that have to be overcome to make this technique work well in a general case. For instance if the pattern to be detected is non-symmetric then consideration must be taken as to how to compare the template in different orientations or to pre-process the image in such a way that the asymmetry is always in the same orientation (Schalkoff, 1989). It is also necessary to consider the effects of scaling on the image under consideration since it is nearly always important to recognise the target pattern regardless of its size.

Many authors have investigated the problems of rotation, scaling and translation invariant systems of template matching. All the systems considered apply some sort of transformation to either the image, the template or both to correct for these effects before attempting the matching process. For instance Yüceer and Oflazer (1993) recommend a method that transforms the image under consideration in order to reduce the number of templates to be considered to two. However they admit that there are some problems to be overcome in the process as a certain amount of distortion is introduced to the image as a result of the transformation process. In this study the Fourier-Mellin transform will be used, this technique produces a rotational, transitional and scale invariant template (for further details see Schalkoff (1989), Turton (1997)).

12.3.1 Discussion of the Fourier-Mellin invariant descriptor

Consider an image $b(x,y)$ which is a rotated, scaled and translated copy of $a(x,y)$,

$$b(x,y) = a(\sigma[\cos\alpha\, x + \sin\alpha\, y]\text{-}x_0,\ \sigma[-\sin\alpha\, x + \cos\alpha\, y]\text{-}y_0)$$

where α is the rotation angle, σ the scale factor, and x_0 and y_0 are the translation offsets.

The Fourier transforms, $B(u,v)$ and $A(u,v)$ of b and a respectively, are related by:

$$B(u,v) = e^{-i\phi_{b(u,v)}}\sigma^{-2}\left|A\sigma^{-1}[u\cos\alpha + v\sin\alpha],\sigma^{-1}[-u\sin\alpha + v\cos\alpha]\right|$$

where $\bullet_{\,b(u,v)}$ is the spectral phase of the image $b(x,y)$. This phase depends on the rotation, translation, and scaling, but the spectral magnitude

$$\left|B(u,v)\right| = \sigma^{-2}\left|A\sigma^{-1}[u\cos\alpha + v\sin\alpha],\sigma^{-1}[-u\sin\alpha + v\cos\alpha]\right| \tag{12.1}$$

is invariant for translations. Equation 12.1 shows that a rotation of image $a(x,y)$ rotates the spectral magnitude by the same angle α and that a scaling of σ scales the spectral magnitude by σ^{-1}. However at the spectral origin ($u=0,v=0$) there is no change to scaling or rotation. Rotation and scaling can thus be decoupled around this spectral origin by defining the spectral magnitudes of a and b in polar co-ordinates (θ,r).

$$A_p(\theta,r) = \left|A(r\cos\theta, r\sin\theta)\right|;0 \le \theta < 2\pi, 0 \le r < \infty$$
$$B_p(\theta,r) = \left|B(r\cos\theta, r\sin\theta)\right|;0 \le \theta < 2\pi, 0 \le r < \infty$$

which leads to

$$B_p(\theta,r) = \sigma^2 A_p(\theta - \alpha, r\sigma)$$

Hence an image rotation (α) shifts the image along the angular axis, and a scale change (σ) is reduced to a scaling along the radial axis and magnifies the intensity by a constant (σ^2). This scaling can be further reduced by using a logarithmic scale for the radial co-ordinate:

$$A_{pl}(\theta,\lambda) = A_p(\theta,r)$$

and

$$B_{pl}(\theta,\lambda) = B_p(\theta,r) = \sigma^2 A_p(\theta-\alpha,\lambda-\rho)$$

where $\lambda = \log(r)$ and $\rho = \log(\sigma)$. This leads to both rotation and scaling now being simple translations, so that taking a Fourier transform of this polar-logarithmic representation reduces these effects to phase shifts, so that the magnitudes of the two images are the same.

This is known as a Fourier-Mellin transform and can be used to compare a single template against an unknown image which will be matched even if it has undergone rotation, scaling or translation.

12.4 PATTERN MATCHING FOR CONCEPT DISCOVERY

Before any patterns of social structure can be considered it is necessary to determine a variable or series of variables available in the 1991 census of population that can be considered to be a proxy for social status. There have been many attempts in the 1980s and 1990s to define what constitutes deprivation both for an individual and for an area, and how best to measure this using variables that are collected in the censuses of population. It is generally recognised that it is important to avoid focusing on groups who are vulnerable to deprivation, but not of themselves deprived. Questions such as, are the deprived always poor or are the poor always deprived? often arise out of this sort of work (Townsend and Gordon, 1991). There has also been considerable work carried out on the links between deprivation and ill health (e.g. Morris and Cairstairs (1991), Jarman (1984)). This study however requires a less specific definition since a proxy for social class is required rather than a specific indicator of deprivation. So for the purposes of this study it was decided to use a combination of local authority rented households, unemployment, overcrowded households (more than one person per room) and households lacking a car and lacking central heating. The variables were normalised so that the highest cell value in each city is 100 and the remainder are scaled relative to the value of that cell.

To generate a pattern of the social structure of the selected urban areas the normalised variables are added together to give a single "deprivation" variable. This can be seen for Leeds in Figure 12.4. The pattern can be clearly seen with the concentration of low social status seen to the north east of the city centre and a smaller centre of low status seen in the Headingly area to the north west of the centre. The 'deprivation' variable surface was then created for each of the 129 urban areas extracted in section 12.3. Once this operation was completed the Fourier-Mellin invariant transform was calculated for each urban area.

Figure 12.4   Social structure of Leeds, see text for details

The next stage in the pattern recognition exercise was to investigate the theoretical patterns of urban social structure discussed above. This was undertaken by creating synthetic patterns that matched the structures that were predicted by theory. Figure 12.5a shows a simple radial decay from high deprivation in the centre to low deprivation at the outskirts, Figure 12.5b shows a stepped radial decay pattern as proposed by Burgess (1925) and Figure 12.5c shows one of the sectoral models used. Sectoral patterns with 4, 6 and 8 sectors were used, as discussed by Hoyt (1939). For each of the synthetic patterns used the Fourier-Mellin invariant transform was calculated.

The next step was to discover which, if any, of the selected urban areas had any similarity with the theoretical models by calculating the root mean square error between the FMI of each urban area and each theoretical pattern. The results of this process were then ranked by level of similarity. It was found that the stepped radial decay model was most similar to any of the urban areas tested, but that the simple linear decay model was also similar to a like group of urban areas. These towns and cities are listed in Table 12.2. The figure in brackets that follows each town name is the level of similarity between that town and the model. The same towns can be seen to match the stepped and linear models of radial decay, although the levels of similarity are better for the linear models compared to the stepped model. This is to be expected due to the similarity of the two theoretical models. However the sectoral model is similar to a very different group of towns and the levels of similarity are much worse than for the radial models.

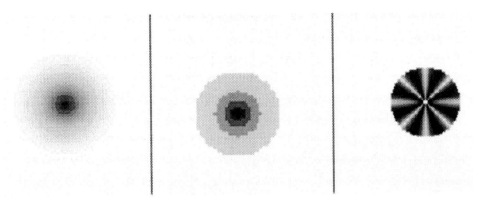

Figure 12.5 Three synthetic cities used for pattern detection: a) Simple radial decay; b) Stepped radial decay; c) Sectoral Model

The towns that are most similar to the radial models tend to be small, inland towns, with the exceptions being Doncaster, Scarbourgh and Corby. It seems likely that the majority of these towns grew slowly around markets before the industrial revolution and have retained their compact shape since. In contrast, the sectoral towns are larger industrial towns and also include a number of seaports (Bristol, Weston-Super-Mare, Hull, Plymouth, Aberdeen and Cardiff). There is a possibility that this was caused by the similarity of a half radial model to a sectoral model. This possibility was tested by calculating the cross correlation of a half radial model with the sectoral models and the radial model. The sectoral model had a correlation coefficient of 9.20 with the half radial model while the radial model had a correlation coefficient of 5.93. If the correlation was spurious it would have been expected that the sectoral model would have been more similar to the half radial model than the radial model was to the half radial model. Therefore the patterns are really sectoral and it is probable that their location on the sea has forced a more sectoral growth pattern since growth is constrained to be away from the sea. Seaports are also more likely to require distinct transport routes to their hinterland.

The remaining industrial towns in the sectoral group are also likely to have developed strong transport links to allow both the import of raw materials and the export of finished goods to the surrounding areas.

To investigate the hypothesis that urban areas in Great Britain show similarity amongst themselves the similarity of each of 129 selected areas and all the other urban areas were calculated. Then for each town or city it is possible to select the other urban areas that are the most similar to this town. From these lists it is possible to determine groups of towns and cities that are more similar to each other than to other groups.

The seven groups with the highest internal similarity are shown in Table 12.3. In each group the largest difference in similarity between two urban areas within the group is 0.5 and in most cases is much smaller. Therefore it can be seen that the similarity levels within these groups is much smaller than the levels reported in the experiment above, which compared theoretical models and urban areas where the lowest level of similarity found was 0.43.

Table 12.2 Towns and cities ranked in order of similarity to theoretical models (see text for discussion)

Stepped Radial Decay Model	Linear Radial Decay Model	Sectoral Model
Skipton (0.43)	Corby (0.48)	Sheffield (3.80) 8
Corby (0.44)	Skipton (0.50)	Leicester (3.28) 8
Harrogate (0.46)	Harrogate (0.54)	Coventry (3.83) 8
Doncaster (0.49)	Doncaster (0.57)	Nottingham (3.97) 6
Blackburn (0.51)	Blackburn (0.59)	Bristol (4.13) 4
Colchester (0.52)	Colchester (0.61)	Weston-Super-Mare (4.23) 4
Scarbourgh (0.56)	Scarbourgh (0.64)	Derby (4.35) 8
Chesterfield (0.57)	Redditch (0.65)	Hull (4.36) 4
Gateshead (0.57)	Gateshead (0.66)	Bury (4.39) 8
Redditch (0.58)	Chesterfield (0.67)	Plymouth (4.40) 6
Wakefield (0.59)	Wakefield (0.67)	Aberdeen (4.42) 6
Macclesfield (0.60)	Macclesfield (0.68)	Cardiff (4.50) 8

Table 12.3 Groups of similar urban areas

Group 1	Group 2	Group 3	Group 4
Airdrie	Motherwell	East Kilbride	Dundee
Woking	Grimsby	Southport	Preston
Macclesfield	Scunthorpe	Bath	Reading
Aylesbury	Stevenage	Accrington	Northampton
Chester	Barrow-in-Furness	Crewe	Newport
High Wycombe	Runcorn	Lowestoft	Petersfield
Corby	Leamington Spa	Bracknell	Ipswich
Scarbourgh	Bedford	Hastings	Wigan
Harrogate	Barnsley	Basingstoke	Lowestoft
Skipton	Chesterfield	Ellesmere Port	
Redditch	St Albans	Mansfield	
Aldershot	Oxford	Halifax	

Group 5	Group 6	Group 7	
York	Aberdeen	Cardiff	
Darlington	Petersfield	Portsmouth	
Hartlepool	Preston	Coventry	
Edinburgh	Warrington	Farnborough	
Gateshead	Reading	Nottingham	
Carlisle	Swansea		
	Northampton		
	Dewsbury		
	Leicester		

12.5 CONCLUSION

The aim of this work was to investigate the potential for the use of computerised generalisation as a solution to the problem of large geographic data sets. It sought to demonstrate that image processing methods which had been developed for the use of other disciplines could be successfully transferred to geography and to investigate the use of computer vision methods in the specific context of the social structure of urban areas.

The first step has demonstrated how, after some suitable pre-processing of the raster data set, the SURPOP population surface could be successfully segmented into urban and rural areas. This led to the extraction of 129 urban areas for Great Britain. The second stage showed how the urban areas of Great Britain could be compared to theoretical models of social structure that were developed predominantly in the United States. It should also be noted that this method of analysis allows the researcher to ignore many of the problems that have occurred in previous statistical studies of social structure, such as 'where is the centre of a city?', 'where are the boundaries of the city?' and 'How wide should the rings in the model be?'

Therefore, in conclusion, it appears that large geographical data sets can be processed by computer to extract simplifications of the data set and to proceed beyond this point to extract new concepts which can lead to new theory formulation with relatively little input from the human researcher. This process could prevent geographers drowning in the sea of new data that is being collected both by industry and new satellites every day.

ACKNOWLEDGEMENTS

This work was partly funded by EPSRC grant GR/K43933. I am grateful to David Henty of EPCC for his help in parallelising the FMI algorithm, and to him and David Martin for their helpful comments on an earlier version of this work. I must also thank Stan Openshaw for his valuable input to early stages of the work. The census data used in this project is Crown copyright and was purchased for academic use by the ESRC and JISC.

REFERENCES

Ballard, D. and Brown, C., 1982, *Computer Vision*. (New Jersey: Prentice Hall).

Bourne, L., 1971, *Internal Structure of the City*. (New York: Oxford University Press).

Bracken, I. and Martin, D., 1989, The generation of spatial population distributions from census centroid data. *Environment and Planning A* Vol. 21, pp. 537-43.

Burgess, E., 1925, The growth of the city. In *The City*, Park, R., Burgess, E. and McKenzie, R. (eds) (Chicago: Chicago University Press), pp. 37-44.

Hoyt, H., 1939, *The structure of Growth of residential neighbourhoods in American Cities*. (Washington).

Hurd, R., 1903, *Principles of City Land Values*. (New York).

Jarman, B., 1984, Underprivileged areas: validation and distribution of scores. *British Medical Journal,* Vol. 289, pp. 1587-1592.

Morris, R. and Cairstairs, V., 1991, Which deprivation? A comparison of selected deprivation indexes. *Journal of Public Health Medicine,* Vol. 13, pp. 318-326.

Niblack, W., 1986, *An introduction to digital image processing* (2nd ed.). (New York: Plenum Press).

Openshaw, S., 1994, A concepts-rich approach to spatial analysis, theory generation, andscientific discovery in GIS using massively parallel computing. In *Innovations in GIS*, Worboys, M. (ed) (London: Taylor and Francis), pp. 123-138.

Schalkoff, R., 1989, *Digital image processing and computer vision*. (New York: Wiley and Sons).

Townsend, P. and Gordon, D., 1991, What is enough? New evidence on poverty allowing the definition of minimum benefit. In *Sociology of Social Security,* Alder, M., Bell, C. and Sinfield, A. (eds) (Edinburgh: Edinburgh University Press).

Turton, I., 1997, *Application of pattern recognition to concept discovery in geography.* M.Sc., School of Geography, University of Leeds.

Yüceer, C. and Oflazer, K., 1993, A rotation, scaling and translation invariant pattern recognition system, *Pattern Recognition* Vol. 26(5), pp. 687-710.

PART IV

Beyond Two Dimensions

13

A critical evaluation of the potential of automated building height extraction from stereo-imagery for land use change detection

Jasmee Jaafar and Gary Priestnall

13.1 INTRODUCTION

Recognising geographic objects in imagery offers much potential for automating land use change detection and for constructing and visualising three-dimensional scenes (Georgopoulos *et al.*, 1997), but it is a hugely complex task. The use of height information extracted from stereo-images is seen to offer a valuable extra clue as part of a wider recognition strategy. The aim of this chapter is to consider the reliability of such height information using a range of surveying methods to provide ground truth.

Several techniques for building extraction from imagery have been proposed. The use of shadow analysis is reported by Irvin and Mckeown (1989). Building detection based on extracting lines from edges or corners and their perceptual grouping has been proposed by Sarkar and Boyer (1993). However, there is no single technique which is sufficiently powerful to solve all the problems in automatic image recognition. The fusion of extra clues or knowledge has been suggested by Shufelt and McKeown (1993) to overcome this single technique restriction.

Integrating prior Geographical Information System (GIS) knowledge towards the automatic recognition and reconstruction of cartographic objects has been examined by a number of researchers. Locherbach (1994) used GIS to aid the reconstruction of the geometry of land use parcels and their subsequent classification. It is clear that existing GIS databases can play a major role in providing the prior knowledge necessary for automatic recognition from imagery. In automated land use change detection, Priestnall and Glover (1998) use vector GIS databases to guide object recognition. This work highlighted the potential of using existing vector databases to 'train' a change detection system and showed how other clues to a building's existence in the image could reinforce this procedure. The addition of building height to such a recognition strategy was see as an important avenue of future research (Murray, 1997; Ridley *et al.,* 1997).

Weidner and Forstner (1995) consider generic object models, where the assumption that buildings are higher than their surrounding surface allows the subtraction of the Digital Elevation Model (DEM) from the 'real' surface, termed the Digital Surface Model (DSM). Regions that may contain buildings are identified as *blobs* and subsequent analysis is restricted to these areas. This subtraction technique has been used by many researchers, however the accuracy of the derived blobs has received little attention. A major research question to be addressed by the current work is the degree to which buildings of a few metres in height can be identified due to the uncertainty involved in the DEM and DSM construction.

The research presented here aims to evaluate the accuracy of the blobs that result from the subtraction of the DEM from the DSM and to assess their potential as an additional clue in a land use change detection methodology.

13.2 METHODOLOGY

The main aim of this chapter is to analyse the accuracy of the blob information derived from the subtraction of the DEM from the DSM. Figure 13.1 shows the procedure of extracting blobs which are potential candidates for the detection of man-made objects. This method has been used in several studies to isolate surface features from the terrain (Baltsavias *et al.*, 1995; Ridley *et al.*, 1997).

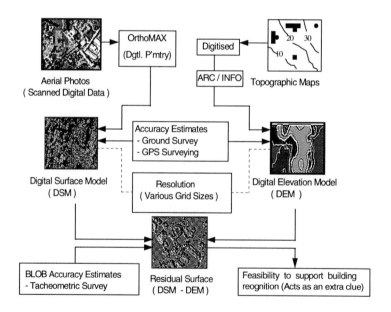

Figure 13.1 The evaluation of the residual surface (DSM - DEM)

For this study, a reference DEM was created within Arc/Info from contour lines derived from Ordnance Survey 1:10 000 map data and spot heights from Ordnance Survey 1:1250 map data. A DSM, which ideally models man-made objects as well as the terrain, was then constructed using Imagine OrthoMax using aerial photography provided by the National Remote Sensing Centre (NRSC). The imagery were 1:10 000 colour aerial photographs of the University of Nottingham campus area, scanned at 21 μm, and having a base:height ratio of 0.6.

The accuracy estimates of the two constructed models were then analysed at various grid resolutions. Based on the assumption that the man-made objects stand above a reference surface, a residual surface was derived by subtracting the DEM from the DSM. Figure 13.2 shows the residual surface and the blobs, which represent the potential candidates for buildings.

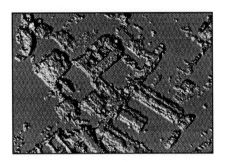

Figure 13.2 Potential man-made objects shown as 'blobs'

13.3 MEASURES OF ACCURACY

DEM and DSM accuracy is of crucial importance as errors in these derived products will propagate through subsequent spatial analysis (Bolstad and Stowe, 1994; Takagi and Shibasaki, 1996; Hunter and Goodchild, 1997) and ultimately affect object recognition. Accuracy describes how close a data value is to truth, relating the value to an established standard reference. It is a relative rather than an absolute concept: it cannot be endowed with a rigid definition such as 'in exact conformity to truth' or 'free from error or defect'. Accuracy can be measured in different ways (Robinson, 1994), both quantitatively (or numerically) and qualitatively (or descriptively). However, the quantitative expression is much more acceptable to the scientific community (Li, 1991).

The most frequently used measure of accuracy is the Root Mean Square Error (RMSE) of the elevational residuals at randomly selected check points. RMSE is calculated from the following equation (Harley, 1975; Gao, 1997):

$$RMSE = \sqrt{\left[\frac{1}{n} \sum_{i=1}^{n} \left(Z_i^* - Z_i \right)^2 \right]}$$

(13.1)

where n is the number of check points; Z_i^* is the terrain elevation at position i and Z_i is the derived elevation at check point i.

Determination of check points to evaluate the derived surface is a crucial stage. Check points at random positions in the study area were established using GPS, traversing and trigonometric levelling techniques with known orders of accuracy. One hundred and seventy nine randomly selected points were established. Table 13.1 shows the corresponding estimated accuracy of the derived check points.

Table 13.1   Accuracy estimates with respect to method employed

Method	Accuracy Estimates		
	X	Y	Z
GPS	0.02 m	0.02 m	0.01 m
Traversing	0.18 m	0.01 m	0.03 m
Trig. Levelling (Reduced Level of Building's Roof)	0.10 m	0.10 m	0.10 m

13.3.1   Accuracy of DEMs

The RMSE of the residuals computed using Equation 13.1 at various grid resolutions are shown in Figure 13.3. The plot shows that the RMSE values change dramatically at a grid resolution of 15 m as described by other researchers (Gao, 1997), this being one of many factors influential in DEM accuracy (Li, 1993). Apart from the grid resolution, the accuracy of the constructed DEM is also affected by other factors. Li (1992) states that the accuracy of the constructed digital elevation model depends on the following factors:

1.   the three attributes (accuracy, density and distribution) of the source data;
2.   the terrain characteristics;
3.   the method used in constructing the DEM; and
4.   the characteristics of the DEM surface constructed from the source data.

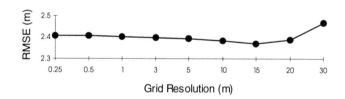

Figure 13.3   RMSE of DEM at various grid resolutions

13.3.2   Accuracy of DSMs

Historically most photogrammetric operations were analogue in nature, involving stereo-plotters which perform measurements on hard copy imagery (Lillesand and Kiefer, 1994). Nowadays, soft copy or digital photogrammetry plays a leading role in acquiring and structuring information from imagery.

The accuracy of the DSMs was evaluated using 70 randomly selected check points. The RMSE plot is shown in Figure 13.4. Figure 13.4 shows, surprisingly, that there is no linear relationship between RMSE and grid resolution between 0.25 m to 5 m. An investigation into this pattern will form an important part of the ongoing research in this study.

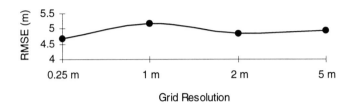

Figure13.4 RMSE of DSM at various grid resolutions

13.3.3 Accuracy of the residual surfaces

The candidate areas for man-made objects are revealed by the residual surface (DSM - DEM) as blobs. DEMs and DSMs have an inherent level of accuracy relating to the grid resolution, and these accuracy levels will eventually be inherited by the residual surface (DSM - DEM).

The accuracy of the residual surface is evaluated at thirty-five random check points. The difference between the heights of the derived blobs (from the residual surface) and true building height are used in determination of the RMSE. Figure 13.5 depicts the RMSE of the residual surface with varying grid resolutions.

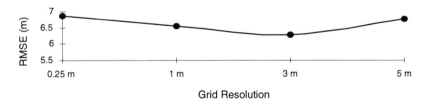

Figure 13.5 RMSE of the residual surfaces

Figure 13.5 shows that the RMSE of the residual surface is between 6 and 7 m. This shows that the uncertainty of the derived blobs is approximately ±7 m from the reference surface. This RMSE is within the extreme limits of the DEMs and DSMs used. Man-made objects which are less than 7 m in height might not be detected if the assumption that all man-made objects are above the reference surface (DEM) is made. Figure 13.6 illustrates the size of the uncertainty zone in relation to a typical profile through the residual surface blobs extracted in this particular urban environment. In certain urban environments, using certain digital data sources, the zone of uncertainty may be dominant over the blobs extracted and therefore less confidence could be placed in the height information when contributing to a recognition strategy.

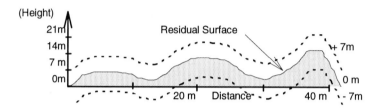

Figure 13.6 Blob uncertainty

Figure 13.7 shows the contour plot of -1 m, 0 m and +1 m level derived from the residual surface. Referring to Figure 13.7, building edges can be detected not only above but below the residual surface at -1 m contour level. Knowing the uncertainty zone, it is clear that possible structure can be identified from interrogation of the residual surface by generating a contour plot at preferred intervals from the lowest uncertainty level (lower bound of the resulting RMSE).

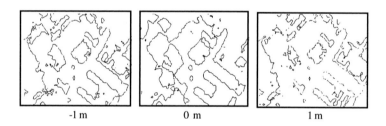

Figure 13.7 Contour plots at -1 m, 0 m and 1 m level

The contouring technique will not only detect possible building edges, but has the potential to separate the detected objects (buildings, trees, etc.) from each other by analysing the contours plot at various levels. Figure 13.8 shows that the buildings become separated from each other at the 6 m contour plot of the residual surface. Characterising different object types based on this technique will be the subject of further research.

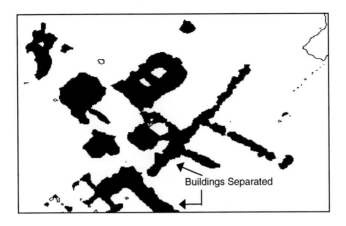

Figure 13.8 Contour plot of the residual surface at 6 m level

13.4 CONCLUSION

In this chapter the possibility of detecting building structures using height information derived from the DSM-DEM subtraction is demonstrated. However, understanding the uncertainty involved with the digital models used is crucial. It is shown that structure may lie below the residual surface due to the accuracy of the digital model being used. This implies that structures of a certain height might not be detected if the assumption that all structures are above the residual surface is made. The results shown suggest that understanding the accuracy of the digital models used is vital if the technique of DSM-DEM subtraction is to be used to detect man-made objects. The contouring technique illustrates that man-made objects can be detected at various levels above and below the residual surface and this offers the potential to characterise and therefore separate objects of different types.

Digital photographs and very high resolution Satellite images will increasingly provide a potential source of height information for man-made objects. The typical height of the objects under study must however be considered in relation to the degree of uncertainty involved with particular data sources if height is to play an important role in an object recognition strategy.

ACKNOWLEDGEMENTS

The authors wish to thank Prof. P.M. Mather for his valuable suggestions, supervision and critical review of the manuscript. The authors wish to acknowledge Dr. Norkhair Ibrahim, Dr. Mohd. Sanusi Ahemad for their constructive comments and Kenny Gibson from the Institute of Engineering Surveying and Space Geodesy, University of Nottingham for his assistance in the GPS survey. Imagery used courtesy of the National Remote Sensing Centre, Barwell, Leicesterchire, UK. Digital Elevation Model based upon Ordnance Survey Data, ED 273554.

REFERENCES

Baltsavias, E., Mason, S. and Stallman, D., 1995, Use of DTMs/DSMs and orthoimages to support building extraction in *Automatic Extraction of Man-Made Objects from Aerial and Space Images*, Gruen, A., Keubler, O. and Agouris, P. (eds) (Basel:Birkhauser Verlag), pp. 199-210.

Bolstad, P. V. and Stowe, T., 1994, An evaluation on DEM accuracy: Elevation, Slope, and Aspect. *Photogrammetric Engineering and Remote Sensing*, Vol. 60(11), pp.1327-1332.

Gao, J., 1997, Resolution and accuracy of terrain representation by grid DEMs at micro scale. *International Journal of Geographical Information Science*, Vol. 11(2), pp. 199-212.

Georgopoulos, A., Nakos, B., Mastoris, D. and Skarlatos, D., 1997, Three-dimensional visualization of the built environment using digital orthography. *Photogrammetric Record*, Vol. 15(90), pp. 913-921.

Harley, J. B., 1975, Ordnance Survey Maps a descriptive manual. (Oxford: University Press).

Hunter, G. J. and Goodchild, M. F., 1997, Modeling the uncertainty of slope and aspect estimates derived from spatial databases. *Geographical Analysis*, Vol. 29(1), pp. 35-49.

Irvin, R. B. and McKeown, D. M., 1989, Methods for Exploiting the Relationship Between Buildings and Their Shadows in Aerial and Space Imagery, *IEEE Transactions On System, Man, and Cybernetics*, Vol. 19(6), pp. 1564-1575.

Li, Z., 1991, Effects of check points on the reliability of DTM accuracy estimates obtained from experimental tests. *Photogrammetric Engineering and Remote Sensing*, Vol. 57(10), pp. 1333-1340.

Li, Z., 1992, Variation of the accuracy of digital terrain models with sampling interval. *Photogrammetric Record*, Vol. 14(79), pp. 113-128.

Li, Z., 1993, Mathematical models of the accuracy of digital terrain model surfaces linearly constructed from square gridded data. *Photogrammetric Record*, Vol. 14(82), pp. 661-674.

Lillesand, T. M. and Kiefer, R. W., 1994, *Remote sensing and image interpretation*, third edition, (New York: John Wiley & Sons).

Locherbach, T., 1994, Reconstruction of land-use units for the integration of GIS and remote sensing data. In *Spatial Information from Digital Photogrammetry and Computer Vision*, Ebner, H., Heipke, C. and Eder, K. (eds) Vol. 30(3), (Munich: ISPRS), pp. 95-112.

Murray, K., 1997, Space imagery in National geospatial database development. In *Proceedings of BNSC seminar on high resolution optical satellite missions* (London: British National Space Centre).

Priestnall, G. and Glover, R., 1998, A control strategy for automated land use detection: An integration of vector-based GIS, remote sensing and pattern recognition. In *Innovations in GIS 5*, Carver, J. (ed) (London: Taylor and Francis), pp. 162-175.

Ridley, H. M., Atkinson, P. M., Aplin, P., Muller, J. P. and Dowman, I., 1997, Evaluating the potential of the forthcoming commercial U.S. high-resolution satellite sensor imagery at the Ordnance Survey. *Photogrammetric Engineering & Remote Sensing*, Vol. 63(8), pp. 997-1005.

Robinson, G. J., 1994, The accuracy of Digital Elevation Models derived from digitized contour data. *Photogrammetric Record*, Vol. 14(83), pp. 805-814.

Sarkar, S. and Boyer, K. L., 1993, Integration, inference, and management of spatial information using Bayesian networks: Perceptual Organization, *IEEE Transactions On Pattern Analysis and Machine Intelligence*, Vol. 15(3), pp. 256-274.

Shufelt, J. and McKeown, D. M., 1993, Fusion of monocular cues to detect man-made structure in aerial imagery, *Computer Vision, Graphics, and Image Processing: Image Understanding*, Vol. 57(3), pp. 307-330.

Takagi, M. and Shibasaki, R. 1996, An interpolation method for continental DEM generation using small scale contour maps. *International Archives of Photogrammetry and Remote Sensing*, XXXI, Part B4, Vienna, pp. 847-852.

Weidner, U. and Forstner, W., 1995, Towards automatic building extraction from high resolution digital elevation models, *ISPRS Journal of Photogrammetry & Remote Sensing*, Vol. 50(4), pp. 38-49.

14

Terrain modelling enhancement for intervisibility analysis

Mark Dorey, Andrew Sparkes, David Kidner,
Christopher Jones and Mark Ware

14.1 INTRODUCTION

Many of today's GIS provide tools that perform terrain analysis. One such tool is the line-of-sight (LOS) function, which calculates the intervisibility between an observer and a target location. There has been much previous work carried out in this area (Fisher, 1991; Franklin and Ray 1994; De Floriani and Magillo, 1994; Wang *et al.*, 1996), leading to a variety of LOS strategies and algorithms. Multiple LOS operations can be performed from a fixed observer to multiple target locations to form a visibility map or viewshed (Yoeli, 1985). These intervisibility functions make use of representations of the Earth's surface, termed digital terrain models (DTMs), which are inevitably subject to error. The quality of a particular DTM will have an obvious effect on the accuracy of any LOS analysis performed on it. It has been shown that small elevation errors can propagate through to large application errors, particularly for intervisibility and viewshed analysis (Huss and Pumer, 1997). It follows that for many applications, very accurate DTMs are required in order to obtain meaningful results. This chapter sets out to raise awareness of some of the error-related problems that can arise when performing intervisibility analysis, and suggests ways in which these can be addressed.

An important issue, when calculating LOS, is the choice of an appropriate surface representation technique. The two most well known DTMs are the digital elevation model (DEM) and the triangulated irregular network (TIN). DEMs, which are made up of a regular grid of elevation values, are by far the most extensively used surface model. This is mainly due to their simplicity and widespread availability. TINs, on the other hand, are not so readily available, and are typically constructed 'in house' from other data sources. In the work presented here, we construct TINs from elevation contours and other surface-specific points. In doing so, we encounter the well-documented problem of 'flat triangles'. A technique is presented that corrects all such anomalies.

Another important issue when performing intervisibility analysis is the selection of a suitable LOS strategy. Calculating the LOS involves interpolating elevation values from discretely surveyed data for a series of unknown sample points. In this paper we demonstrate how the location and density of sample points, together with the choice of interpolation algorithm, can have an effect on the LOS result. It is noted that most GIS

169

fail to document their implementation of these procedures. As a result, the user is unable to place a measure of confidence on the results of any analysis (Fisher, 1993).

Furthermore, it is often the case that DTMs are supplied as the ground elevations only, and lack any topographic features such as the heights and geometry of buildings and vegetation. We propose that the quality of a DTM, in terms of LOS accuracy, can be increased significantly by including within it digital representations of real-world features. With this in mind, we put forward methodologies for improving DTM quality by incorporating topographic features into both DEMs and TINs.

Finally, we provide details of a validation tool using the Virtual Reality Modelling Language (VRML) to identify inconsistencies between our digital models and the real world. VRML is defined by a public specification document (VRML, 1997) and is rapidly becoming a favoured 3D modelling language due to its accessibility, portability and modest hardware requirements. The approach provides a means whereby inconsistencies in the surface model or topographic features can be identified. The model may then be edited using the information gained from the VRML representation to more accurately reflect the landscape it represents.

The work presented here has been verified by a fieldwork study centred on the Taff Ely wind farm in South Wales. The visibility of wind turbines is a contentious issue, which often leads to many proposed wind farm developments being rejected at the planning stage (Kidner *et al.*, 1998). Their size, and dominance within the landscape, make them an ideal candidate on which to conduct visibility analysis. To this end, several hundred observer locations were identified and measured within a 10 km radius of the wind farm, using a differential GPS (Global Positioning System), to accurately record each *X*, *Y* and *Z* co-ordinate. Validation at a number of benchmarks suggested that the sampled GPS elevations were accurate to within a few centimetres. At each GPS location, digital photographs were taken to record the view and thus provide a means of verification with our models and VRML representations.

This chapter is arranged as follows. First, in Section 14.2, we address the issue of surface model choice and methods for improvement. In Section 14.3 we discuss various LOS strategies and interpolation algorithms. Next, in Section 14.4, we put forward methodologies for improving DTM quality by incorporating topographic features into both DEMs and TINs. Section 14.5 provides details of the VRML validation tool that is used to identify inconsistencies between our models and the real world. The paper concludes in Section 14.6 with discussion and some preliminary results from our fieldwork study.

14.2 CONSTRUCTION AND IMPROVEMENT OF SURFACE MODELS

The Ordnance Survey (O.S.) provides complete national coverage for the U.K. of digital contour data from surveyed scales of 1:50,000 (20 x 20 km tiles) and 1:10,000 (5 x 5 km tiles) at 10 metre and 5 metre contour intervals respectively. The O.S. derives DEMs from these products at resolutions of 50 metres and 10 metres respectively.

Contour data can also be used to create TINs by the application of algorithms such as a constrained Delaunay triangulation. A common problem when creating TINs from digital contours is the possibility of forming flat regions in the DTM which do not exist in the actual terrain. These are termed invalid flat triangles. However, not all flat regions are invalid as the terrain may contain features such as lakes and plateaux. Invalid flat

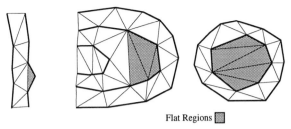

Flat Regions ▦

Figure 14.1 Creation of flat triangles when triangulating contour lines

triangles often occur along contours when the distance between points is less than the distance between separate contours (Figure 14.1). Where the Delaunay triangulation algorithm finds a point's two Delaunay neighbours (that will thus form a single triangle) lying on the same contour, a flat triangle will be formed. Removing flat triangles from a triangulation is a non-trivial task and yet is necessary in order to create a DTM that more accurately represents the terrain on which it is based.

A common preventative solution is to remove points from the contour lines by an automatic process known as line simplification as suggested by E.S.R.I. (1991). A widely used algorithm for achieving this is Douglas and Peucker's (1973) line generalisation algorithm. This process seeks to eliminate superfluous points along the contour whilst still maintaining its fundamental shape. A disadvantage of this technique is that it fails to take into account those nodes that define surface specific points, thus important terrain information may be lost. The altered lines may also take on a more linear form as the generalisation proceeds. This may lead to contours crossing, resulting in topological errors in the final triangulation.

Although generalising contours reduces the number of flat triangles, it will not eliminate them completely. A number of algorithms have been proposed to solve this problem (Brandli, 1992; Vozenilek, 1994). These algorithms typically involve the swapping of triangle edges and/or the insertion of points subsequent to the initial triangulation of the contour data. To this end, we have implemented our own algorithm for removing flat triangles. A summary of the algorithm follows, full details of which are described by Ware (1998).

The algorithm begins by sequentially searching through the triangulation identifying

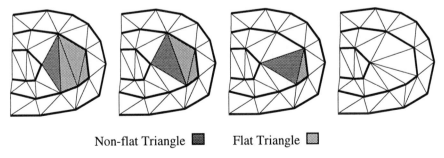

Non-flat Triangle ▦   Flat Triangle ▦

Figure 14.2 Resolving flat triangles between two contours by
swapping edges of a flat triangle and its non-flat neighbour

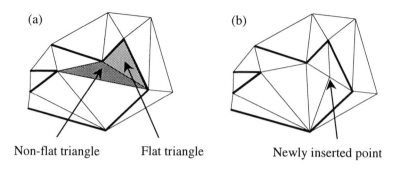

Non-flat triangle   Flat triangle

Newly inserted point

Figure 14.3 Resolving a flat triangle when an angle inside the quadrilateral
formed by a non-flat and flat triangle, is greater than 180°

flat triangles. When an invalid flat triangle is discovered, the algorithm proceeds by searching for all adjacent flat triangles, thus forming a contiguous flat region. A search path is stopped if either a contour edge or a non-flat triangle is found. If a non-flat triangle has been found, then the flat region lies between a pair of contours, otherwise it lies within a single contour. If the former is true, then the algorithm attempts to correct the flat region by a series of edge swaps (Figure 14.2). An edge swap is permitted only if the quadrilateral formed by the current flat triangle and the current non-flat triangle is convex (i.e. all internal angles are less than 180 degrees). Having swapped an edge the procedure always ensures that the next non-flat triangle is the newly created triangle with three non-contour edges. The process continues until either all the invalid flat triangles that are in the current flat region have been corrected, or a quadrilateral is formed that is not convex. In the latter case, a new point is inserted into the triangulation, at the midpoint of the shared edge (Figure 14.3). Once a new point has been inserted, the algorithm returns to

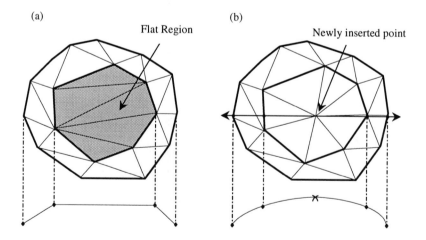

Figure 14.4 Resolving an invalid flat region by the insertion of a new point. Its height is derived by
fitting a polynomial curve through surrounding contour edges

the main search through the triangulation where any flat triangles remaining in the current flat region will be corrected later. If a non-flat triangle is not found that borders the flat region (i.e. it lies within a single contour line), then the procedure inserts a new point at its centre before re-triangulating the region (Figure 14.4). The height of this new point is calculated by sending out rays in different directions and checking to see which contour edges are intersected (Clarke *et al.*, 1982). Having found this information the gradient in each direction is checked to find the most dominant slope at that point. The final elevation is calculated by fitting a polynomial curve through the elevation points in the chosen direction (Figure 14.4b). Whilst these procedures are all automatic, care should be taken to ensure that the algorithm does not try to remove valid flat regions.

14.3 LOS STRATEGIES AND INTERPOLATION ALGORITHMS

Once the choice of the surface model has been made, there are a number of factors to consider when calculating the LOS that can have a significant bearing on the visibility result (Fisher, 1993). Two such factors are the sampling interval and the interpolation algorithm used.

14.3.1 Selection of sampling strategy

There are a number of DEM sampling strategies. One such approach is to sample wherever the path of the LOS intersects a pair of DEM vertices or 'grid lines' i.e. between row, column and diagonal neighbours (Edwards and Durkin, 1969). Other strategies may include random or uniform sampling along the LOS.

A number of algorithms exist in the literature for computing LOSs on TINs (Goodchild and Lee, 1989; Lee, 1991; De Floriani and Magillo, 1994). Many of these algorithms compute predicted visible polygons by dividing individual triangle facets into visible and non-visible areas, albeit at the expense of algorithm complexity and thus computational time. A simple method of calculating the LOS on a linear based TIN is to use the 'brute force' or 'ray-shooting' algorithm (De Floriani and Magillo, 1994). This technique traverses the LOS over the TIN surface, checking at each intersected triangle edge to see whether the elevation angle is greater than that of the LOS angle (Figure 14.5). Where this is the case, the LOS is blocked.

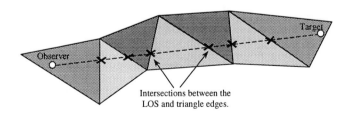

Intersections between the
LOS and triangle edges.

Figure 14.5 Determining LOS on a linear based TIN using a ray-shooting algorithm.

14.3.2 Choice of interpolation algorithm

The interpolation algorithm used to calculate unknown elevations in the DTM is often critical to LOS analysis. In the case of DEMs, algorithms range from stepped algorithms, where each grid cell is assigned a uniform elevation thus forming a series of plateaux, to more complex polynomial surfaces which use the surrounding elevations of the individual grid cell, to ensure continuity and possibly smoothness. For example, consider the regular interpolation of elevations within the shaded grid cell of the discretely sampled vertices of Figure 14.6. When visualised as a surface view from the south west, Figure 14.7 illustrates a number of different interpolation scenarios, which in this example results in a variance in interpolated height of up to 18 metres at the same locations.

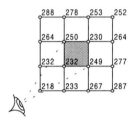

Figure 14.6   Grid cell viewed from SW direction.

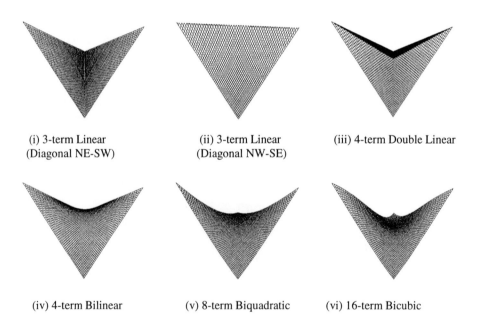

(i) 3-term Linear
(Diagonal NE-SW)

(ii) 3-term Linear
(Diagonal NW-SE)

(iii) 4-term Double Linear

(iv) 4-term Bilinear

(v) 8-term Biquadratic

(vi) 16-term Bicubic

Figure 14.7   Interpolation results using various interpolation algorithms.

In the case of TINs, a number of interpolation algorithms exist, ranging from simple linear interpolation at triangle edges to more complex algorithms, such as quintic and cubic interpolation. The more elaborate techniques use surrounding triangle nodes to estimate slopes or partial derivatives, which are used to enforce smoothness. In all cases, the use of more complex algorithms incurs greater computational effort

14.4 MODELLING OF TOPOGRAPHIC FEATURES

Having addressed the relevant issues with regard to surface models, LOS strategy and interpolation algorithms, we now examine the inclusion of topographic features into our surface models. Many DTM data sources do not include topographic features for example buildings and vegetation, instead they simply model the underlying terrain. In an attempt to represent the landscape more accurately, 3D topographic features should be incorporated into the terrain model. This data can be obtained from a number of sources, including cadastral mapping, aerial photographs and paper maps. Having incorporated features from a number of different sources (and possibly different scales), it should be noted that the accuracy of individual features, in terms of their relative sizes and positions, might compromise the outcome of any analyses performed. Also, many data sources do not include specific feature attribute information, such as its height, and therefore, it must be derived by other means.

The assignment of attributes can be a time-consuming process, particularly when many features are involved. In the case of buildings for instance, only their height, type and roof-structure may need to be stored. However, when incorporating vegetation other attributes might possibly be required. These could include information such as its type, for example shrub, hedgerow, oak tree or beech tree. Having assigned a type, the height can then be attributed. This information may be known or else have to be estimated. Additionally, temporal information such as whether the feature is coniferous or deciduous (and therefore affected seasonally), the initial date of observation, its growth pattern (including maximum height), might also be recorded. Using this temporal information the possibility exists for the model to evolve over time. For example, hedgerows are often cropped in autumn, whereas other vegetation may be left to grow naturally until it reaches maturity. Factors such as these play a significant part in affecting the outcome of a predicted LOS and therefore should be considered when performing visibility analysis.

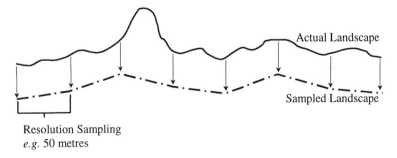

Resolution Sampling
e.g. 50 metres

Figure 14.8 Effect of having a large sampling resolution on a DEM.

14.4.1 Representing topographic features in DEMs

One of the drawbacks in using a DEM is its inability to model rapidly changing terrain due to the regular sampling of its elevation vertices (Figure 14.8). This drawback has implications when trying to model topographic features, as these may also represent abrupt changes in surface 'complexity.' A common method of modelling a topographic feature is to overlay it onto the DEM. The elevations of any vertices covered by the feature are then increased accordingly by the feature's height (Figure 14.9a). This has a number of disadvantages. Firstly, the true size and shape of an individual feature may be lost, as wherever any 'overhang' exists the remainder of the object is merely disregarded (Figure 14.9b). Secondly, features that are enclosed within an individual grid square, such as a tree or hedge, are unable to be modelled, as they do not cover grid vertices (Figure 14.9c). Lastly, vertical faces are unable to be modelled. Reducing the sample resolution to a size small enough to register abrupt changes in height over small distances can approximate this. Thus, in order to approximate the vertical face of a building, the DEM resolution must be very small, which consequently leads to a large increase both in storage costs and data redundancy. If the DEM resolution is not small enough, then sloping edges are formed between vertices (Figure 14.10a).

If vertical faces are not modelled correctly the accuracy of any visibility analysis performed on the DEM is likely to be compromised. Figure 14.10b shows how a 10 metre high building might be represented on a 10 metre DEM. The generation of the sloping edge wrongly increases both the size and area occupied by the topographic feature. If every topographic feature is modelled in this way, the slopes may have a significant effect on the accuracy of a LOS, and ultimately on the final viewshed. This can be avoided by not explicitly incorporating topographic features into the model, but rather by storing them separately in either a vector or raster format, for which a brief description follows.

(a)   (b)   (c)

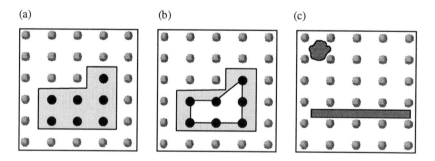

Figure 14.9 Topographic features in DEMs: (a) DEM nodes covered by building polygon; (b) DEM representation of polygon; (c) Inability of DEM to model polygons enclosed within single grid squares.

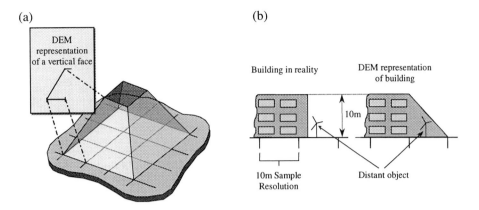

Figure 14.10 Incorporation of a topographic feature into a DEM: (a) A feature covering just one grid square will affect the surrounding squares and (b) the possible effect of sloping edges

14.4.1.1 Vector representation of data

In the case of vector data, an initial LOS test is performed on the DEM without topographic features being considered to examine if the terrain obstructs the LOS. If the LOS is unobstructed, a line intersection algorithm is used to determine the locations of any intersections between the LOS profile and the topographic feature polygons. To minimise search times, these polygons may be indexed with, for example, a secondary grid, quadtree or R-tree. Having identified any intersections on a given topographic feature polygon, the terrain elevation is interpolated and stored, together with its feature type, for example, building or vegetation. The feature type information is necessary to determine how the feature is to be placed in the landscape. For example, the top of a 3 metre high hedgerow will always be the same height above the landscape regardless of undulations in the terrain beneath it (Figure 14.11a). Therefore, when modelling vegetation, its height is simply added to that of the stored interpolated elevation at each intersection point. However, a different technique is needed for buildings whose foundations typically lie flat in or on the terrain, to ensure horizontal floors and ceilings (Figure 14.11b). Therefore, the maximum, minimum or average terrain elevations at all polygon intersection points are first calculated. The feature's actual height is then added to this value, resulting in a flat polygon foundation. However, there are instances where the roofs of terraced buildings follow the landscape in much the same way as vegetation, and should be modelled as such.

Having assigned the polygon a suitable height, the angle from the observer location to each intersection point is calculated and compared to the LOS angle between the observer and the target. If the LOS angle is smaller, then the LOS is blocked. The advantages of this methodology are twofold. Firstly, vertical faces are represented correctly. Secondly, topographic features such as walls and fences which are typically represented as lines rather than polygons, can be directly included in the vector data model.

(a)   (b)

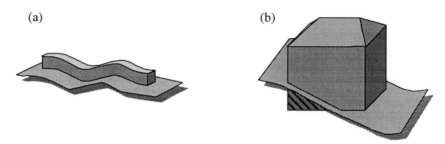

Figure 14.11   Placement of topographic features in the DTM: (a) Vegetation follows undulations of
the terrain and (b) a building with flat foundations irrespective of underlying terrain.

14.4.1.2   *Raster representation of data*

Topographic features can be stored as raster data in a separate layer, covering the same
area as that of the DEM. When performing intervisibility analysis, each interpolated point
is checked to see if it falls within a rasterised pixel. Where this is the case, the feature is
identified and its height is added to the interpolated base elevation from the DEM. A
problem can arise in the original rasterising process where inaccuracies occur because the
boundaries of individual raster cells (pixels) rarely conform to the boundaries of
polygons. The vector-to-raster algorithm therefore tends to shift the original polygon
boundaries to coincide with the boundaries of individual raster cells, thus distorting their
shape. These distortions may be quite significant, particularly where the raster cell size is
large in relation to the size of the polygon boundary features (Veregin, 1989). The rules
for deciding whether or not a pixel belongs to a polygon are varied. Some algorithms
might use *dominant unit rasterising* where the majority of the pixel i.e. greater than 50%,
must be covered by the vector polygon in order for it to be rasterised. Others such as
central point rasterising state that if the centre of the pixel lies inside the polygon then it
is rasterised (Knaap, 1992). As a general rule, the smaller the raster pixel, the better the
representation of the vector polygon, albeit at the expense of greater storage
requirements. For example, a pixel size of 1 metre may produce acceptable raster results,
whereas a 10 metre pixel size will reduce storage overheads, but at the expense that the
feature may not be accurately represented, and in some cases, missed out completely
(Figure 14.12). The storage requirements for a typical 4 km^2 area are 75 kB and 7,500 kB
for 10m and 1m sampling respectively.

14.4.2   **Representing topographic features in TINs**

An often-cited advantage of the TIN data model over the DEM is its ability to model
sudden changes in the terrain by the inclusion of non-crossing breaklines and exclusion
boundaries, thus forming a constrained Delaunay triangulation. This capability can also
be used to incorporate topographic features into the TIN. To include a feature, two sets of
constraints are incorporated into the raw contour data. The first is the feature outline; the
second is created by marginally shrinking the feature so that it forms an inner boundary
inside the original (Figure 14.13). Having inserted the objects into the contour data each
one must be given an elevation that is in keeping with the height of the terrain at

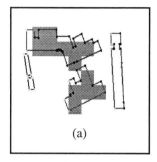 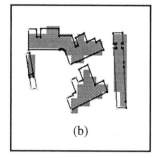 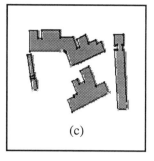

(a)	(b)	(c)

Figure 14.12 Effect of raster pixel size on feature representation accuracy using:
(a)10 m, (b) 5 m, and (c) 1 m pixels.

its location. Two possibilities exist. Firstly, that the object intersects one or more existing contour lines and secondly, that the object lies between two contours. In the first case, the object's base height is taken as either the highest, lowest or average intersected contour elevation (Figure 14.14a). If no intersection takes place then a series of checks are made to find the heights and locations of surrounding contours. This is achieved by sending out rays from the object in different directions and checking to see which contours are intersected (Clarke *et al.*, 1982). Having found this information the gradient in each direction is checked to find the most dominant slope at that point. The final elevation is calculated by fitting a polynomial curve through the elevation points in the chosen direction (Figures 14.14b and 14.4b). Having determined the feature's base elevation, the internal constraints are assigned a height equal to that of the feature plus the base elevation. These constraints force the triangulation to create triangles that are near vertical, depending on the precision which the user finds acceptable, thus allowing objects such as buildings and hedgerows to be represented accurately (Figure 14.15).

This methodology can be extended further to model more complex structures, i.e. the roofs of buildings. In the real world the variety of roof-structures that exist means that precisely modelling each building's roof is often impractical. However a generic

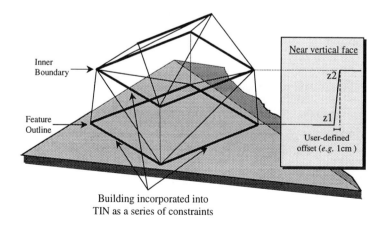

Figure 14.13 Incorporation of a topographic feature into a TIN.

approach that produces more realistic roof-scapes than those of simple flat roofs, is to generate the skeleton (or medial axis) of each building polygon. These are then inserted into the TIN, as constraints and assigned heights equal to their respective feature. The heights of the internal wall constraints are then reduced accordingly in order to create the roof pitch.

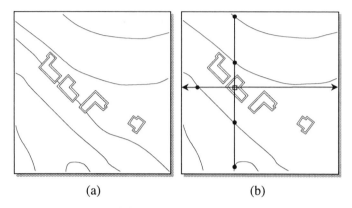

(a)   (b)

Figure 14.14 Calculating the height of an object when it
(a) intersects a contour line and (b) lies in-between two contours.

14.5 LOS VALIDATION

Although the techniques previously described improve the models' representations, the process of validation is useful in order to identify further disparities, which have not as yet been resolved. These may include errors such as missing topographic features, incorrect attribute data of those features (i.e. height information, locational discrepancies), or even elevation errors in the underlying base model itself. Correct interpretation of terrain data is often assisted by the use of 2.5 and 3D visualisation tools, for example Arc/Info and ArcView 3D-Analyst. The recent introduction of VRML provides potential for the efficient interactive rendering of a variety of surface models, including DTMs. We describe here how a VRML implementation can aid in the validation of predicted LOSs.

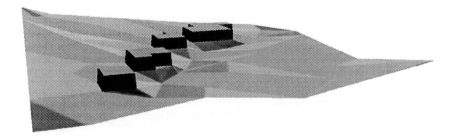

Figure 14.15 Incorporating buildings into a TIN. Note the near vetical faces of the walls.

The main benefits of using VRML as a validation tool for visibility on DTMs are: its ability to give an exact 3D representation of the data models; its flexibility in allowing many viewpoints to be visited and viewed in a semi-realistic manner; its ease of use in allowing navigation through the virtual landscape; its versatility in defining data types due to its programming functionality; and its ability to deal with multiple instances of predefined objects, for example, different building types such as houses, churches, and schools.

In order to visualise accurately the terrain in VRML, it is important that the representation precisely relates to the data structure of the DTMs. In respect to DEMs, VRML provides the 'surface elevation' construct, which takes the grid dimensions and vertex elevations as parameters in order to construct the virtual terrain model. In contrast, the terrain as described by the TIN consists of a series of non-overlapping contiguous triangular facets of differing dimensions. These can be directly modelled by specifying each facet as a VRML 'Indexed Face'.

Topographic features can easily be incorporated into a VRML world either by constructing individual flat planes or else by defining them using VRML's key geometric shapes. In the case of the DEM, each feature is modelled separately and placed onto the terrain, its base elevation having been interpolated beforehand. This method provides the flexibility of being able to texture map, or colour each object in order to add realism and definition. This technique may also be used on the TIN. The disadvantage of this approach is that it utilises additional polygon data as well as the basic DEM data structure. This in turn may compromise the accuracy of the VRML model as it is unlikely to exactly correspond with the original model as used by the LOS algorithm. As mentioned previously, the TIN has the advantage over the DEM in that it allows topographic features to be directly incorporated into the DTM structure itself. Thus the resulting VRML representation can be made to model the TIN data precisely, with each topographic feature being directly 'stamped' into the landscape.

Digital photographs were taken facing the wind farm to record the view at each registered GPS point. This was done in order to provide a means of identifying deficiencies in our existing data models, such as missing topographic features, incorrect heights of existing features, incorrectly assigned vegetation types, inaccurate modelling of specific features (e.g. roof-structures), and inaccurate terrain representation. Figure 14.16 shows a typical example where a number of factors contribute to the generation of an erroneous computed view. Having identified any inconsistencies, rectification can then be attempted in order to model more closely the actual view and thus hopefully, achieve a better correlation between the predicted and actual visual impact of the wind farm at that location.

A visual impact assessment should include all the relevant information about the proposed new development. An important part of this is the predicted visual impact on the local environment. This is commonly shown using computer generated Zones of Visual Intrusion (ZVIs) or viewsheds (Kidner *et al.*, 1996). The viewshed consists of areas of visible terrain from an observer's viewpoint within that terrain. Profiles that radiate out from the observer are plotted as LOS profiles showing the visible and non-visible areas. The visible areas in each LOS profile are joined to adjacent visible areas in other LOS profiles to form the complete viewshed (Yoeli, 1985). A disadvantage of this method is that although it gives an indication of whether a location is visible or not, it fails to convey how the view from that location is likely to be affected by the new development. VRML provides the possibility of allowing locations to be specified

Figure 14.16 Comparison of photograph with VRML representation. Note the difference in the amount of terrain visible above the houses, the lack of vegetation, and the incorrect modelling of roof-structures

(perhaps via a 2D map or a street address), before being presented with likely 3D impressions of those views. Thus a more interactive and intuitive means of understanding the likely visual impact from individual locations is possible. If the data used to construct the VRML representation is the same as that for the terrain on which the viewshed was calculated, then there should be a good correlation between the impression given by the 3D view and the results of the viewshed analysis.

14.6 OBSERVATIONS AND CONCLUSIONS

In many respects, visibility analysis is still very much in its infancy. At present, there has been little verification of the accuracy of visibility results produced by many GIS, although some work has been carried out by Fisher (1993). In his work, Fisher compares several GIS visibility algorithms using the same data. His results showed significant discrepancies even to the extent of some algorithms depicting non-visible areas within the visible areas produced by others. The conclusions drawn were that in many cases, the visibility algorithms implemented in commercial GIS are not sufficiently documented, and hence, their reliability cannot be evaluated.

The debate between TINs and DEMs has reached a point where the benefits and disadvantages of both data models are well documented and understood (Ware and Kidner, 1996). Most GIS use the DEM data structure for visibility analysis due to its relative simplicity, high availability and ease of use when performing many spatial operations. Whilst TINs have been widely acknowledged as being a better means of surface representation, they have largely been the favoured model of academics, rather than commercial or mainstream GIS users. This may be because TINs require more complicated algorithms to manipulate the explicitly defined topology (Theobald, 1989). In practice, many GIS require that a TIN be first converted to a DEM before performing certain operations, including visibility analysis. However, this casual approach to terrain modelling is likely to increase the chance of introducing application errors.

Much of today's DTM data does not contain any topographic feature information, but just records the underlying terrain elevations. As a result, many analyses will be intrinsically flawed due to the incomplete nature of the information incorporated within the DTM. This situation is set to change with the increased availability of DTM products generated using fine resolution remote sensing techniques, such as laser-scanning. This technology is able to produce high-resolution data models that accurately reflect both the terrain and its features (Flood and Gutelius, 1997).

Improving the quality of the DTM by correcting terrain deficiencies, incorporating topographic features and verifying the LOSs against known viewpoints provide a means of increasing the reliability and confidence levels of LOS analysis on DEMs and TINs. In our case study of the Taff Ely wind farm several hundred observer locations have so far been analysed. Of these, approximately 60% were located in built-up areas, with the remainder in open countryside. Initial results suggest that the accuracy of prediction using the techniques detailed above, as opposed to the base elevation DTM, can increase by as much as 77%. Future work will involve analysing further GPS observer locations, some of which will be up to a radius of 20 km away from the wind farm. Other work will include probable LOS analysis to provide confidence levels on individual LOS predictions. This work will provide further information on the accuracy and performance of our models over varying distances, together with a reliability measure of the results obtained.

ACKNOWLEDGEMENTS

The authors would like to acknowledge the collaborative and financial support of National Wind Power throughout the course of this research.

REFERENCES

Brandli, M., 1992, A triangulation-based method for geomorphological surface interpolation from contour lines. In *Proceedings of the 3rd European Conference and Exhibition on GIS (EGIS '92)*, (Utrecht: EGIS Foundation), 1, pp. 691-700.

De Floriani, L., and Magillo, P., 1994, Visibility algorithms on triangulated digital terrain models. *International Journal of Geographical Information Systems,* Vol. 8(1), pp. 13-41.

Clarke, A. L., Gruen, A. and Loon J. C., 1982, The Application of Contour Data for Generating High Fidelity Grid Digital Elevation Models. In *Proceedings of the Fifth International Symposium on Computer-Assisted Cartography* (Auto-Carto 5), (Bethesda, Maryland: ACSM/ASPRS), pp. 213-222.

Douglas, D. H. and Peucker, T. K., 1973, Algorithms for the reduction of the number of points required to represent a digitised line or its caricature. *The Canadian Cartographer,* Vol. 10(2), pp. 112-122.

Edwards, R. and Durkin, J., 1969, Computer prediction of service areas for V.H.F. mobile radio networks. In *Proceedings of IEE,* Vol. 16(9), pp. 1493-1500.

E.R.S.I., 1991, Surface modelling with TIN: surface analysis and display, *ARC/INFO User's Guide.*

Fisher, P. F., 1991, First experiments in viewshed uncertainty: the accuracy of the viewshed area. *Photogrammetric Engineering & Remote Sensing,* Vol. 57(10), pp. 1321-1327.

Fisher, P. F., 1993, Algorithm and implementation uncertainty in viewshed analysis. *International Journal of Geographical Information Systems,* Vol. 7(4), pp. 331-337.

Flood, M. and Gutelius, B., 1997, Commercial implications of topographic terrain mapping using scanning airborne laser radar. *Photogrammetric Engineering & Remote Sensing,* Vol. 63(4), pp. 327-366.

Franklin, W. R. and Ray, C., 1994, Higher isn't necessarily better: visibility algorithms and experiments. In *Advances in GIS,* Waugh, T. (ed) (London: Taylor & Francis), pp. 751-770.

Goodchild, M. F. and Lee, J., 1989, Coverage problems and visibility regions on topographic surfaces. *Annals of Operations Research,* Vol. 18, pp. 175-186.

Huss, R. E. and Pumer, M. A., 1997, Effect of database errors on intervisibility estimation. *Photogrammetric Engineering & Remote Sensing,* Vol. 63(4), pp. 415–424.

Kidner, D. B., Dorey, M. I. and Sparkes, A. J., 1996, GIS and visual assessment for landscape planning. In *Proceedings GISRUK 96,* University of Kent, Canterbury, pp. 89-95.

Kidner, D. B., Sparkes, A. J. and Dorey, M. I., 1998, GIS in wind farm planning. In: *Geographical Information and Planning: European Perspective,* Geertman, S., Openshaw, S. and Stillwell, J. (eds) (Springer-Verlag) *in press.*

Knaap, W. G. van der, 1992, The vector to raster conversion: (mis)use in geographical information systems. *International Journal of Geographical Information Systems,* Vol. 6(2), pp. 159-170.

Lee, J., 1991, Analysis of visibility sites on topographic surfaces. *International Journal of Geographical Information Systems,* Vol. 5(4), pp. 413-429.

Theobald, D. M., 1989, Accuracy and bias issues in surface representation. *Accuracy of Spatial Databases,* Goodchild, M. and Gopal, S. (eds) (London: Taylor & Francis), pp. 99-105.

Veregin, H., 1989, Review of error models for vector to raster conversion. *Operational Geographer,* Vol. 7(1), pp. 11-15.

Vozenilek, V., 1994, Generating surface models using elevations digitised from topographical maps. In *Proceedings of the 5th European Conference and Exhibition on GIS (EGIS '94),* (Utrecht: EGIS Foundation), 1, pp. 972-982.

VRML, 1997, http://www.vislab.usyd.edu.au/vislab/vrml/specs.html

Wang, J., Robinson, G. J. and White, K., 1996, A fast solution to local viewshed computation using grid-based digital elevation models. *Photogrammetric Engineering & Remote Sensing,* Vol. 62(10), pp. 1157-1164.

Ware, J. M., 1998, A procedure for automatically correcting invalid flat triangles occurring in triangulated contour data. *Computers and Geosciences,* Vol. 24(2), pp. 141-150.

Ware, J. M. and Kidner, D. B., 1997, A flexible storage efficient TIN data model. In *Joint European Conference and Exhibition on GIS,* (Amsterdam: IOS Press), pp. 48-57.

Yoeli, P., 1985, The making of intervisibility maps with computer and plotter. *Cartographica,* Vol. 22(3), pp. 88-103.

15

Distributed viewshed analysis for planning applications

Philip Rallings, David Kidner and Andrew Ware

15.1 INTRODUCTION

Intervisibility and viewshed analysis are amongst the most common functions of those GIS that support digital terrain modelling. In the main, they are used to aid planning decisions, such as the siting of contentious developments with the aim of minimising visual intrusion, or alternatively for identifying locations which maximise the field-of-view, such as for broadcast coverages, scenic viewpoints, watchtowers, or missile defences (Franklin and Ray, 1994). Whilst the viewshed quantifies visibility for a limited set of test locations, its major shortcoming is that it does not identify optimal locations. An alternative is the reverse viewshed, which quantifies the visibility at each possible location within the planning zone, thus identifying both intrusive and discreet sites. However, whilst viewshed analysis is widely acknowledged as being processor-intensive, the calculation of the reverse viewshed can increase this workload by many more orders-of-magnitude. This chapter describes the implementation of the reverse viewshed algorithm on a distributed cluster of networked machines. The practicality of parallelising many GIS operations is now within the reach of many users, such as corporate industry, academia, local and national government, and environmental agencies, without the need for any further investment. The key to unlocking this opportunity is to demonstrate how existing resources (i.e. networked computers) can be used for popular, but processor-intensive tasks.

15.2 VISUAL IMPACT ANALYSIS

There is an increasing requirement for many of today's environmental impact assessments (EIAs) to consider the problem of visual intrusion. EIA is an aid to decision making and a medium by which the environmental effects of a project are conveyed to the public. Its strength is in enabling the best environmental fit between a project and its surroundings, and in helping to determine whether the development is acceptable (I.E.A. and T.L.I., 1995). Whilst landscape impact relates to changes in the fabric, character and quality of the landscape as a result of development, visual impact relates solely to changes in available views of the landscape, and the effects of those changes on people. In predicting visual impacts, the main requirements are to show: the extent of potential/theoretical visibility; the views and viewers affected; the degree of visual intrusion or obstruction that will occur; the distance of the view; and the resulting impacts upon the character and quality of views (IEA and TLI, 1995). For some planning applications, visual intrusion will be the key factor in determining the success or failure of a project. However, visual intrusion is very subjective by nature, as what might be a 'blot on the landscape' to one observer might appear in harmony with the landscape to another. The ability to quantify and qualify visual impacts, scenic values and visual intrusion has been the focus of much research during the last twenty years (e.g. Baldwin (1998), Bishop and Hulse (1994), Carlson (1977), Guldmann (1979), Miller (1995)). In the meantime, visual intrusion is often incorrectly inferred from the EIA by means of the viewshed or intervisibility map and a number of photomontages from selected viewpoints (Figure 15.1). Photomontages are a popular visualisation technique, with the advantage that they show the development within the real landscape and from known viewpoints. Sheppard (1989) provides a good overview of the considerations, techniques and impediments to generating such visual simulations.

Figure 15.1 The photomontage process: original photograph, CAD representation, and photomontage overlay

Intervisibility analysis is a long established function of many geographical information systems (GIS). In the simplest case, it is used to determine whether a target location can be seen from an observer location, termed the line-of-sight or LOS. This can be extended to consider the visible region or viewshed of an observer location, determined as a series of LOS calculations to all other locations on the surface (or vertices of the digital terrain model) within a pre-defined Zone of Visual Influence (ZVI). Alternatively, the viewshed can be thought of as the regions from which the subject of the planning application can be seen. The binary viewshed can also be extended to consider multiple observer (or target) locations, in which the number of unobstructed LOSs is summed. This is commonly known as the cumulative viewshed (Figure 15.2).

(a)

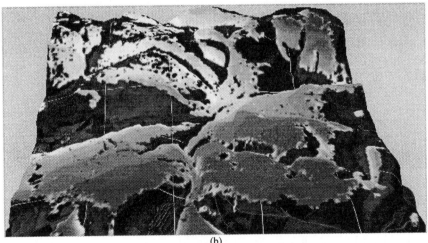

(b)

Figure 15.2 The cumulative viewshed of the 20 wind turbines for the Taff Ely wind farm in South Wales (a) orthophotos draped on the 5 x 5 km O.S. 1:10,000 scale DEM with the turbines identified as white markers (b) the cumulative viewshed, where the grey scale colour ramp ranges from white (1) to dark grey (20)

The cumulative viewshed is the traditional approach chosen by many developers to portray the visibility of developments such as wind farms in the EIA. For such complex developments, the alternative is to produce a sequence of maps that portray regions from which all of the development will be visible through to partly affected regions. The cumulative viewshed of Figure 15.2 is illustrated below in Figure 15.3 to identify individual households from which the wind farm is visible and an associated quantitative measure in terms of how many wind turbines are visible.

Figure 15.3   The cumulative viewshed of Figure 15.2 visualised to identify the quantitative impact of turbine visibility with respect to households

The preceding chapter (Dorey *et al.*, 1999) outlines many of the issues involved with respect to improving the accuracy of intervisibility analysis for planning applications. The guidelines for landscape and visual impact assessment (I.E.A. and T.L.I., 1995) suggest that any analysis should acknowledge any deficiencies or limitations inherent in the assessment. This is clearly not adhered to when viewshed or cumulative viewshed maps are presented in the EIA, often to the detriment of the developer submitting the proposal. The effects of elevation errors, topographic features such as buildings and vegetation, and algorithm errors have been largely ignored in EIAs to date. Fisher (1996) details a variety of viewshed functions and presents a methodology for handling uncertainty in these operations. The result is termed the probable viewshed, which should give more reliable responses to visual impact queries.

15.3   THE REVERSE VIEWSHED

With respect to planning, the viewshed function gives some indication of visibility for a pre-defined location or set of locations. When considered with other elements of the landscape or environmental assessment, the viewshed function is very constrained and inflexible. Developers and planners like to consider 'what if' scenarios. For example, 'what if this group of wind turbines were sited 100 metres further back?', or 'what needs to be done to mitigate the number of visible turbines on this housing estate?'. The viewshed quantifies visibility on the surface in terms of 'times seen', but does not qualify the results to answer such queries.

One simple extension to the viewshed function is to consider the reverse viewshed as the cumulative viewshed or visibility index of each location within the proposed site. For our example of the wind farm, rather than consider the 20 or so pre-determined turbine locations, the reverse viewshed might consider thousands of possible locations, usually at regular intervals, say 10 metres or so. The reverse viewshed gives an indication of the site areas sensitive to visibility, and possibly visual impact. For example, consider the 22 x 22 km DTM of South Wales centred on the Taff Ely wind farm (Figure 15.4). The DTM is a regular grid digital elevation model derived by the Ordnance Survey from the 1:50,000 scale Panorama series, sampled at 50 metre intervals. The reverse viewshed is calculated for each vertex within a 2 x 2 km subset of this DTM, representing the site of the existing wind farm (Figure 15.5). In terms of the total number of lines-of-sight which need to be analysed for this problem, the workload approaches 327 million.

Figure 15.4   Shaded relief map of the 22 x 22 km 1:50,000 scale DTM for South Wales with roads centred on the Taff Ely wind farm (identified by the 2 x 2 km inner white square)

The results of this operation are illustrated in Figure 15.6, where the visibility indices range from 1% (white) to 40% (dark grey) of all the 441 x 441 vertices of the surface model. This example assumes that line-of-sight is defined as from an observer height of 40.6 metres (i.e. at the hub of the possible wind turbine) to a target height of 1.7 metres (i.e. eye level on the ground). Alternative reverse viewsheds can be calculated to other objects of the GIS, such as building vertices or centroids, postcode centroids, road vertices, or DTM vertices weighted by landscape value scores. In all cases, the reverse viewshed can be used to identify locations that can minimise the visual impact of the development. In the example of Figure 15.6, the visibility of the wind farm, in terms of visible area, could have been reduced if more of the turbines in the darker regions had been sited in the lighter regions.

Figure 15.5 The 2 x 2 km Taff Ely site for which the reverse viewshed indices are to be calculated The 20 existing turbine locations are clearly identified

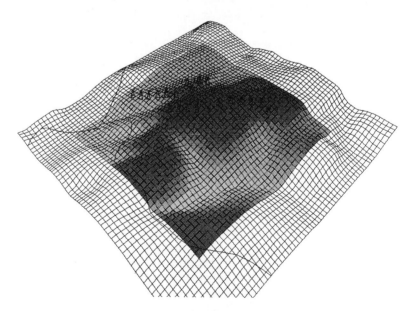

Figure 15.6 The reverse viewshed for the 2 x 2 km Taff Ely site of Figure 15.5 applied to the 22 x 22 km DTM of Figure 15.4. Darker regions indicate areas of higher visibility (up to 40% of the DTM). The actual turbine locations are also shown

15.4 PARALLEL PROCESSING AND GIS

This extension of the viewshed problem into a planning tool inevitably means that uni-processor systems provide poor response times when calculating the visibility indices of more than just a few locations. The problem is compounded if we wish to consider other time-consuming factors, such as the incorporation of algorithms to detect topographic features, higher order interpolation techniques, finer resolution DTMs, or probability modelling (Dorey *et al.*, 1999; Fisher, 1996). In these circumstances, parallel processing techniques can be applied to such computationally intensive GIS applications (Healey *et al.*, 1998).

The recognition that parallel processing has a role to play in the development of GIS has been well documented over the last ten years, particularly with respect to digital terrain modelling applications and intervisibility analysis (De Floriani *et al.*, 1994; Kidner *et al.*, 1997; Magillo and Puppo, 1998; Mills *et al.*, 1992; Teng *et al.*, 1993). However, there is a reluctance to take this technology forward beyond the realms of academia. Gittings *et al.* (1993) indicate that GIS technology, software and functionality have developed rapidly along with user expectations in recent years, to the extent that the available software systems must continue to meet these expectations. This is becoming increasingly more difficult, with the danger that expectations may exceed actual requirements or the capabilities of current technology (Calkins and Obermeyer, 1991). Parallel processing can provide a flexible and cost-effective means of overcoming this potential handicap by extending processing capacity, although considerable effort may sometimes be required to design appropriate algorithms that can exploit these architectures (Gittings *et al.*, 1993). Much of the early research on parallel GIS algorithms focused on using fine grain architectures or Transputer networks. Today, the increasing availability of computer networks in the workplace, combined with recent advances in modern PC operating systems means that many organisations already have an existing multi-purpose parallel processing resource which could be utilised to carry out such tasks.

Networks are commonplace, and when combined with one of the many fully multi-tasking and pre-emptive operating systems, it is possible to build a parallel processing cluster of machines. Operating systems which are able to support these environments include Windows NT, OS/2, LINUX, Solaris, and to a limited extent, Windows 95. These parallel processing clusters can be homogeneous (same machine architecture) or heterogeneous in nature and can link machines or workstations of various speeds and abilities. Each machine in a cluster can be a complete system in its own right, usable for a wide range of other computing tasks. Specialised hardware (parallel or otherwise) can also be incorporated into the cluster or virtual machine.

The main advantage of using a cluster is that many organisations have a network of fast processor machines (e.g. Pentium PC-based machines). Whilst each single system cannot on its own perform the whole task, they are able to process part of the overall job. Cluster computing can scale to provide a very large parallel machine and specialised hardware can be made available to all machines (e.g. a machine in the cluster could forward data to an image processing workstation so that the data can be mapped in to a visual form). Each individual machine would also have total and independent control of its own resources (e.g. memory, disk, etc.) which can also be exploited (e.g. a machine typically has more memory than processors in a more specialised computer environment). Clusters thus enable an organisation to realise the full potential of a computer network, and the combined processing power can handle even the most extensive GIS application.

General-purpose network hardware is not designed for parallel processing systems, especially when masses of data have to be passed between systems. Latency is very high and bandwidth is relatively low compared to specialised parallel processing environments. However, recent developments in network technology make it possible for workstations to have either a higher shared or a dedicated bandwidth (e.g. the cluster shares a 100 mbs bandwidth or each machine has its own dedicated 10 mbs bandwidth). Message passing and communication is the main overhead when using a parallel cluster and is an issue that must be addressed when implemented into GIS applications. Furthermore, if a heterogeneous network is created, some message encoding system must be incorporated so that all machines can communicate effectively, regardless of processor architecture or operating system.

There are a number of readily available software packages for assisting the implementation of distributed systems. These include PVM (Geist *et al.*, 1994), Express (Flower *et al.*, 1991), the Message Passing Interface (MPI Forum, 1994), and LINDA (Carriero and Gelernter, 1989). Various other systems with similar capabilities are also in existence - Turcotte (1993) provides an exhaustive list and overview. To be of practical use, the software must support a number of different vendors and architectures, unless of course, only a homogenous cluster is created.

15.5 PARALLEL VIEWSHED ALGORITHM

One of the main drawbacks to the wider use of parallel processing architectures is the difficulty of detecting the inherent parallelism in problems. Parallel processing involves either splitting the process into several sub-processes and performing them on different processors concurrently, or splitting the data that is to be processed between a number of processors and executing multiple copies of the process simultaneously. Healey and Desa (1989) categorise algorithm parallelism as being one of three broad approaches, namely event, geometric and algorithmic parallelism. Event parallelism, or farming, utilises the concept of a master processor distributing tasks to slave processors and assembles the computational results from each in turn. Geometric, or domain parallelism partitions the data set over the available processors, whilst algorithmic, or function parallelism partitions the process into sub-tasks. East (1995) also provides a good overview of these different approaches and considers the communication issues and overheads of each.

Before considering the approach undertaken, it is worth highlighting the steps involved in determining the reverse viewshed for a specific visibility region (Figure 15.7). For this simplified problem, it can be seen that the number of distinct tasks is relatively few. The processing overheads are due to the sheer quantity of data to be analysed. The main variable in this algorithm that will seriously affect the processing time is the calculation of the LOS profile. In the simplest case of a profile retrieved from the DTM via bilinear interpolation without considering topographic features, the average time taken would be less than one millisecond. In the most extreme case of a profile retrieved from the DTM via higher order interpolation (e.g. a 36-term biquintic polynomial), with consideration for buildings and vegetation, and calculated with an error model as a probable LOS, the average time taken could be a few seconds. For the problem illustrated earlier of 327 million LOS calculations, the latter approach would take more than 10 years to complete. In reality, bilinear interpolation with incorporation of topographic features is more than sufficient.

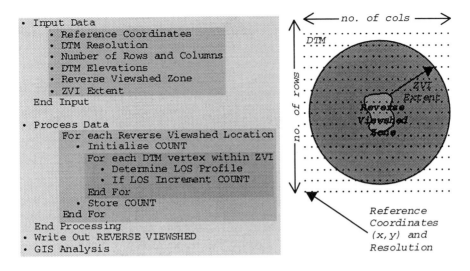

Figure 15.7 Serial algorithm for determining the reverse viewshed

Taking the simplicity of the serial algorithm into account, it becomes apparent that the best approach to parallelising the function is to consider decomposing the domain into manageable workloads for each processor. Each processor needs to be assigned an equivalent amount of work in order to minimise idleness and thus produce a load-balanced system. The load-balancing issue is even more exaggerated given the nature of a heterogeneous network connecting processors of different speeds and abilities. In this respect, the equivalent amount of work would be measured in time rather than volume, with quicker processors being assigned a greater amount of work. The assignment process would be influenced by some measure of machine performance, when compared with others or a base machine. This would enable faster resources to be exploited to their full potential. Another consideration of domain decomposition is overcoming or minimising the problem of accessing data from neighbouring sub-regions. Healey and Desa (1989) note that as individual sub-regions are handled by different processors, boundary data must be either duplicated (overlap) or passed between processors. It is therefore important to define sub-regions, such that the amount of within region processing is maximised and the between region communication is minimised.

15.6 DOMAIN DECOMPOSITION STRATEGIES

Domain decomposition can be achieved by either partitioning the DTM or partitioning the vertices of the reverse viewshed. In the first case, each processor calculates the visibility indices for all of the vertices in the reverse viewshed to a subset of the DTM, whilst in the latter case, each processor calculates the visibility indices for a subset of the reverse viewshed to all of the DTM.

The partitioning of raster data such as DTMs for parallel algorithms requires methods for reading data from disk, distribution of data amongst the processors, gathering the resultant data from the processors, and writing data to disk (Mineter, 1998).

A number of alternative approaches exist, each of which imposes a different pattern of coordination and communication between processes. The partitioning can be achieved either statically or dynamically.

Static allocation involves dividing the DTM between the processors at the start of the process, such that the work allocation does not alter during runtime. The scale of the domain allocated per processor is called the grain size or granularity. The work dividing algorithm would have to allocate the domain evenly amongst all available processors, but if there is no *a priori* knowledge of the processing load then the domain allocation would be one of dividing the workload into equal areas. Static domain decomposition requires regularity in the structure of the domain, uniformity of the workforce requirements, and uniformity of the working capabilities of the workforce. Any non-uniformity in any one of the aforementioned areas may render the decomposition of the domain in such a way as not to be able to preserve processor efficiency. Static allocation is well suited to environments with identical or equally matched processors, however careful attention must be given to the work dividing algorithm such that the work is actually divided into equal amounts of processing time. If certain areas require more processing due to terrain specific features, then the problem of load balancing becomes an even greater issue when preserving overall system efficiency.

Dynamic allocation involves allocating workload (or domain segments) to processors throughout the entire system runtime. Typically worker processors are assigned tasks by a master processor, whilst upon completion, the workers return any results and are subsequently assigned more tasks. Such a process is commonly termed *farming* or collectively known as *processor farms* and is a practical way of implementing dynamic allocation. With respect to Healey and Desa's (1989) and East's (1995) categorisation, the approach is a hybrid of event and domain parallelism. Dynamic allocation is needed when any one of the following is true: the domain structure is irregular and does not match workforce structure; work is distributed non-uniformly across the domain, and work capability is distributed non-uniformly to the workforce. Dynamic allocation is capable of supporting a diverse range of processors and is able to utilise individual capabilities and speeds. The disadvantage of farming is the management overhead, mainly associated with having dedicated control nodes. The additional dynamic management incurs extra communication that is not inherent in the problem and exists solely to support dynamic allocation of the domain. The communication overhead further increases as additional nodes are added to the processor farm, which can produce a communication bottleneck at the control (or farmer) node. The only way to alleviate this is to apply recursion by introducing more control nodes to form a management hierarchy. Dynamic allocation is well suited for heterogeneous networks of varying speed processors and in applications where the workload (or domain) cannot easily be allocated according to processing time. However, the extra management overhead can be minimised to provide a better utilisation of processors. One of the main concerns in domain decomposition is the granularity of the workload allocated to each processor, since it can have a dramatic effect on processor efficiency and management overhead.

A variety of static and dynamic allocation algorithms were implemented for both segmenting the DTM and the reverse viewshed vertices. For the static algorithms, these included pre-assigning processors to equal numbers of rows of the DTM, equal sized blocks (of rows), and equal sized rectangular regions. The segmentation of the reverse viewshed indices was also based on equal numbers of points or blocks. For the dynamic algorithms, segmentation of the DTM was accomplished by individual rows or assigning n vertices per processor, whilst the reverse viewshed vertices were segmented either individually or into small groups of points. For all algorithms, the complete DTM was made available to each processor, as the intermediate terrain between the reverse viewshed vertex and the target location is required to retrieve the LOS profile.

15.7 TESTS AND RESULTS

The parallel visibility algorithms were applied to the problem of determining the reverse viewshed of a 2 x 2 km visibility region of a larger 22 x 22 km regular grid DTM of South Wales (the ZVI). The region is the location of the Taff Ely Wind Farm, 15 km north-west of Cardiff, and is the subject of an on-going investigation into the performance of GIS tools for the visibility analysis of Wind Farms (Dorey *et al.*, 1999). The DTM was sampled at a resolution of 50 m, hence the indices were calculated for 1,681 points (i.e. 41 x 41 vertices) as the number of line-of-sight profiles to every other DTM vertex (194,481 points, i.e. 441 x 441 vertices). This is equivalent to nearly 327 million profile calculations, each of which could consist of up to 1,000 interpolations.

The parallel processing cluster was established in stages for twenty-four 100 Mhz Pentium based machines, connected via a standard 10 mbs ethernet network, running Windows NT Workstation 4.0. The software was developed using the PVM (Parallel Virtual Machine) system and the C programming language. PVM can be compiled on over 40 platforms including specialised and non-specialised parallel processing hardware, including JavaPVM for distributed processing over the Internet. PVM is ideal for generating code, since applications are portable over the PVM system, regardless of machine architecture.

Two comparison indices (Agrawal and Pathak, 1986) were observed for each work-dividing algorithm: speed-up and processor efficiency. Speed-up relates to the time taken to compute a task on an uni-processor system to the time taken to compute the same problem using the parallel implementation. Speed up is thus defined as:

$$Speed\text{-}Up = \frac{\textit{Elapsed Time of a Single Processor}}{\textit{Elapsed Time of the Multi-Processors}}$$

The relative efficiency, based on the performance of the problem for one processor, can be a useful measure as to what percentage of processor time is used on computation. Any implementation penalty would be immediately reflected by a decrease in efficiency:

$$\textit{Efficiency} = \frac{\textit{Speed-Up} \ x \ 100\%}{\textit{Number of Processors}}$$

In terms of comparison, the relative efficiency indicator is the major factor since it has an implied time element in terms of implementation overheads and shorter execution. The efficiency factor also gives an indication as to the effective load balancing of the partitioning algorithm. The graph below (Figure 15.8) presents a selection of the many static and dynamic partitioning algorithms implemented.

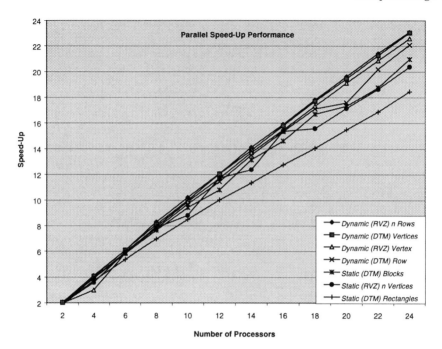

Figure 15.8 Graph illustrating the speed-up in performance with number of processors for a selection of the static and dynamic partitioning algorithms
(RVZ refers to a partitioning of the reverse viewshed zone, whilst DTM refers to a partitioning of the DTM)

For the static algorithms, most notable was the apparent ineffectiveness of algorithms that partition the workload in specific regions of the DTM, such as in rectangles or blocks. Alternative static approaches that break down the processing into smaller workloads in more diverse regions proved to be extremely erratic in terms of efficiency. The main problem is being able to load balance the processors according to equivalent workloads, which is very dependent on the characteristics of the terrain. A region that is not LOS to the reverse viewshed zone because of an intermediate hill or mountain will require little processing time. This needs to be recognised when the workloads are assigned.

Overall the dynamic partitioning approaches produced better speed-up performances than the static algorithms and more consistent efficiency indicators suggesting a well balanced distributed system. The dynamic approach produced some efficiency performances of over 100% compared to the uni-processor system. This is in part due to the different structure of the farming program. In view of this, it may be possible to restructure the serial system to achieve a better performance. The results demonstrated the ineffectiveness of farming with extremes of granularity (too high or low), suggesting a need to observe the optimal granularity for dynamic farming algorithms. From the results it was deduced that optimal efficiency is attained when the granularity represents approximately 0.1% to 0.3% of the domain, since the efficiency drops when the granularity is 0.5% to 1%, and is totally inefficient when the granularity is over 5%. Inefficient results were also observed when the granularity is less then 0.05% of the domain.

The communication overhead involved in managing dynamic farming was quite low, thus demonstrating the effectiveness of the implementations. However, the overhead increases as processors are added, so at some point it will become inefficient, although the results are promising even on 24 processors. The communication overhead (Annaratone *et al.*, 1989) is measured as a percentage of the overall parallel implementation time (not the uni-processor time) and is a necessary burden to ensure that the system is load balanced, with the main factor being the granularity of work allocated. Even on the inefficient algorithms the communication overhead is not excessive, proving that the major drawback is load balancing.

15.8   CONCLUSION

The results prove that load balancing is difficult with static decomposition and can be erratic in relation to the number of processors. Dynamic farming strategies, on the other hand, produced better results and are far more resilient in relation to the processor network and load balancing - this being true when there is no extreme level of granularity.

The algorithms have been tested on a network of homogeneous processors of the same speed, but the problems of implementing static algorithms would be even more pronounced when heterogeneous processors of varying speeds were added to the network. These problems would increase depending on the varying degrees of machine capabilities. Static algorithms are less efficient as more processors are added, even of the same type, and would deteriorate further in a heterogeneous network.

Dynamic processing farms would not suffer from the same penalties, as every processor could be utilised due to the inherent load balancing properties of farming. However, it may be the case that in heterogeneous networks, the granularity may prove to be more of a determining factor than in homogeneous networks, where there is more freedom in terms of granularity.

Currently only a one tier system has been investigated in which one master node communicates with all processing nodes in a tree or hierarchical fashion, and all sub-processes are created and communicate with the master process. In addition, it is currently assumed that all processors have access to the entire DTM. The multi-layered model (Dandamundi and Eager, 1990) would allow the DTM to be divided between additional control nodes such that each sub-system could apply the work dividing algorithms to its own local domain. This model would be capable of processing much larger DTMs, at larger resolutions and still present a viable, well structured distributed GIS. The additional control nodes would certainly be beneficial to the dynamic farming algorithms, even if each processor continued to have access to the entire DTM. Our research suggests that whilst one master node is sufficient for 24 processors, the management communication overhead would spiral as extra processors are added. It is doubtful that a single tree system could cope with many more additional nodes for implementing the dynamic algorithms.

Parallel processing of GIS problems is still in its infancy, despite the steady stream of academic papers that have addressed the issues over the last 10 to 15 years. The need for specialised hardware was certainly a major obstacle to its acceptance by a wider community outside of academia. However, as most organisations now use networked machines, parallel processing has the potential to benefit a much wider audience. The majority of applications that could benefit from parallelisation, such as visibility analysis, are conceptually very simple. The real problem is identifying the best strategy for assigning the workloads to machines, which minimises both processor redundancy and communication overheads. For the development of the reverse viewshed algorithm, which has the potential to be a beneficial tool for developers and planners alike, this chapter has demonstrated approaches which can process the data up to 23 times faster using only 24 processors. It remains to be seen whether the GIS community will adopt parallel computing, but the technology and hardware is now within our reach.

REFERENCES

Agrawal, D. P. and Pathak, G. C., 1986, Evaluating the performance of microcomputer configurations. *IEEE Computer*, Vol. 19, pp. 23-27.

Annaratone, M., Pommerell, C., and Ruhl, R., 1989, Interprocessor communication speed and performance in distributed-memory parallel processors. In *16th Symp. on Computer Architectures*, pp. 315-324.

Baldwin, J., 1998, Applying GIS to determine mystery in landscapes. In *Innovations in GIS 5*, Carver, S. (ed) (London: Taylor and Francis), pp. 179-186.

Bishop, I. D. and Hulse, D. W., 1994, Prediction of scenic beauty using mapped data and geographic information systems. *Landscape and Urban Planning*, Vol. 30, pp. 59-70.

Calkins, H. W. and Obermeyer, N. J., 1991, Taxonomy for surveying the use and value of geographical information. *International Journal of GIS*, Vol. 5(3), pp. 341-351.

Carlson, A. A., 1977, On the possibility of quantifying scenic beauty. *Landscape Planning*, Vol. 4, pp. 131-172.

Carriero, N. and Gelernter, D., 1989, LINDA in Context. *Communications of the ACM*, Vol. 32, pp. 444-458.

Dandamundi, S. P. and Eager, D. L., 1990, Hierarchical inter-connection networks for multicomputer systems. *IEEE Trans. Computers*, Vol. 29(6), pp. 786-797.

De Floriani, L., Montani, C. and Scopigno, R., 1994, Parallelizing visibility computations on triangulated terrains. *International Journal of GIS*, Vol. 8(6), pp. 515-531.

Dorey, M., Sparkes, A., Kidner, D., Jones, C. and Ware, M., 1999, Terrain modelling enhancement for intervisibility analysis. In *Innovations in GIS 6*, Gittings, B. (ed) (London: Taylor and Francis), this volume.

East, I., 1995, *Parallel processing with communicating process architecture*. (London: UCL Press), ISBN 1-85728-239-6.

Fisher, P. F., 1996, Reconsideration of the viewshed function in terrain modelling. *Geographical Systems*, Vol. 3, pp. 33-58.

Flower, J., Kolawa, A. and Bharadwaj, S., 1991, The express way to distributed processing, *Supercomputing Review*, May, pp. 54-55.

Franklin, W. R. and Ray, C. K., 1994, Higher isn't necessarily better: visibility algorithms and experiments. In *Advances in GIS Research* (Proceedings of 6th Int. Symp. On Spatial Data Handling), Waugh, T.C. and Healey, R. G. (eds) (London: Taylor and Francis), pp. 751-770.

Geist, A., Beguelin, A., Dongarra, J., Jiang W., Manchek, R. and Sunderam, V., 1994, *PVM: Parallel virtual machine - a user's guide and tutorial for networked parallel computing*.

Gittings, B. M., Sloan, T. M., Healey, R. G., Dowers, S. and Waugh, T. C., 1993, Meeting expectations: a review of GIS performance issues. In *Geographical*

Information Handling – Research and Applications, Mather, P.M. (ed) (Chichester: John Wiley and Sons), pp. 33-45.

Guldmann, J.-M., 1979, Visual impact and the location of activities: a combinatorial optimization methodology. *Socio-Economic Planning Sciences*, Vol. 13, pp.47-70.

Healey, R. G. and Desa, G. B., 1989, Transputer-based parallel processing for GIS analysis: problems and potentialities, *Proc. of Auto-Carto 9*, ACSM/ASPRS, pp. 90-99.

Healey, R., Dowers, S., Gittings, B. and Mineter, M. (eds), 1998, *Parallel Processing Algorithms for GIS*. (London: Taylor and Francis).

I.E.A. and T.L.I., 1995, *Guidelines for Landscape and Visual Impact Assessment*. (London: E and FN Spon), ISBN 0-419-20380-X.

Kidner, D. B., Rallings, P. J. and Ware, J. A., 1997, Parallel processing for terrain analysis in GIS: visibility as a case study. *GeoInformatica*, Vol. 1(2), pp. 183-207.

Magillo, P. and Puppo, E., 1998, Algorithms for Parallel Terrain Modelling and Visualisation. In *Parallel Processing Algorithms for GIS*, Healey, R., Dowers, S., Gittings, B. and Mineter, M. (eds) (London: Taylor and Francis), pp. 351-386.

Mills, K., Fox, G., and Heimbach, R., 1992, Implementing an intervisibility analysis model on a parallel computing system. *Computers and Geosciences*, Vol. 18(8), pp.1047-1054.

Miller, D. R., 1995, Categorisation of terrain views. In *Innovation in GIS 2*, Fisher, P. (ed) (London: Taylor and Francis), pp. 215-221.

Mineter, M. J., 1998, Partitioning Raster Data. In *Parallel Processing Algorithms for GIS*, Healey, R., Dowers, S., Gittings, B. and Mineter, M. (eds) (London: Taylor and Francis), pp. 215-230.

MPI Forum, 1994, MPI: A message-passing interface standard. *International Journal of Supercomputing Applications*, Vol. 8(3/4).

Sheppard, S.R.J., 1989, *Visual Simulation – A User's Guide for Architects, Engineers, and Planners*, (New York: Van Nostrand Reinhold), ISBN 0-442-27827-6.

Strand, E.J., 1992, Multiple processors promise powerful GIS advancements. *GIS Europe*, May, pp. 34-36.

Teng, Y. A., DeMenthon, D. and Davis, L. S., 1993, Region-to-region visibility analysis using data parallel machines. *Concurrency: Practice and Experience*, Vol. 5(5), pp. 379-406.

Turcotte, L., 1993, *A survey of software environments for exploiting networked computing resources*. Technical Report, Mississippi State University.

16

Four-Dimensional Virtual Reality GIS (4D VRGIS): Research guidelines

Nathan A. Williams

With the continuing growth of interest in the use of Virtual Reality technology in GIS, urban planning and the geosciences, it is becoming clear that the development of a VRGIS will be required. This chapter attempts to outline the challenges faced by the development of such a technology and suggests the principles that should guide future development of four-dimensional VRGIS (4D VRGIS). Secondly, a definition of what 4D VRGIS should be is proposed, and an outline conceptual architecture suggested, with reference to the issues of space, time, co-ordinates, measurement, functionality and structure within a 4D VRGIS.

16.1 INTRODUCTION: VR AND GIS TO VRGIS

During the last three years Virtual Reality (VR) technology has begun to be utilised in research and development into multi-dimensional Geographic Information Systems, (GIS). A survey by Haklay (1998) concluded that this research has concentrated on six areas related to GIS: Urban planning, environmental planning and impact assessment, visualisation, archaeological modelling, education, and military simulation and intelligence. This work has tended to implement high-end solutions, with the simulation of terrains and urban environments being a common theme. Since 1997, however, work has begun to concentrate on developing integrated VRGIS that incorporate a limited amount of common GIS functionality, on desktop computers (Neves *et al.,* 1997; Raper *et al.,* 1997). This integration has been possible for two reasons: The movement of GIS to component technology, and the mainstream acceptance and availability of low cost, but extremely powerful 3D/VR technology. By using component-based GIS and VR technology, systems can be developed that contain both GIS and VR functionality. Such a system could be referred to as a first generation Virtual Reality Geographic Information System or VRGIS.

Work in VR and GIS before 1997 could be described as the 'how can we use this?' approach, i.e. VR as an 'add on' to GIS. GIS were used to generate data views from query and analysis, and VR used to visualise and provide only high-level interaction with the

GIS output. The first generation of component based VRGIS developed after 1997 are a progression on from this approach. However, they still do not address the opportunity for the true integration at a low level because these systems tend to lack even basic GIS functionality in the VR component of the application itself. What is required is a coherent VRGIS Application Programming Interface (API) and appropriate data storage architecture. Such an interface would provide opportunities for developers to utilise the components of geospatial modelling: object types, relationships (spatial/spatio temporal), attributes, conventions and operations, and the full functionality of current VR technology. Thus it is suggested that GIS functionality should be integrated into future Virtual Reality engines. Then we would see a VR tool that is not simply meant as an add-on to existing GIS, but an API and data storage architecture that is a fully functional, integrated development paradigm that would enable the development of true Multi-Dimensional Virtual Reality GIS.

16.2 DEFINING VR AND GIS: ASSESSING THE DIFFERENCES

Before tackling the definition of four-dimensional VRGIS, it is necessary to first examine the basic definitions of VR and GIS. GIS can be defined as an organised collection of computer hardware, software, geographic data, and personnel designed to efficiently capture, store, update, manipulate, analyse, and display all forms of geographically referenced information (ESRI, 1994). Virtual Reality can, in its simplest incarnation, be defined as an immersive, interactive simulation of realistic or imaginary environments. (The term Virtual Reality (VR) was first coined by Jaron Lanier in 1988, when referring to a computer generated, 3D interactive environment.)

At a high level, in some instances both VR and GIS are designed to display representations of reality and abstract data at varying degrees of abstraction, principally with the aim of enhancing user cognition of data and its context. It could thus be argued that in their essence both VR and GIS are similar. However, at lower levels the differences in the inherent architectures of VR and GIS become apparent. The foundation of all VR systems, almost without exception, is the concept of the virtual universe or scene graph. A scene graph is a hierarchical arrangement of nodes that represents objects, their attributes, and position in three or four dimensions, within the virtual universe or simulation. VR universe co-ordinates are based around an arbitrary centre point of 0,0,0, rather than those of the real world in GIS, and do not cater for standard geographic co-ordinate systems. As a rule, VR systems do not provide the facility for geo-referencing as a standard, yet object co-ordinates within a VR simulation can be assigned with universe co-ordinates that parallel the real world. Most conventional GIS on the other hand tend to organise information into two-dimensional layers of a fixed scale and generalisation, with the obvious exception of object-oriented (OO) GIS and 3D GIS. Further to this, the in-built functionality of each technology is quite different. Most VR systems are designed for display and high level interaction (manipulation, navigation and simple database query). GI systems on the other hand include this functionality, (though navigation is usually on a 2D plane through the manipulation of layers, or rotation and zooming in 3D, with display being in non immersive 2/3D), and also include the ability to perform spatial operations upon the data, a feature currently lacking in VR systems.

Although these discrepancies are not discussed in depth here, further development must be guided by research that defines the exact suitability of VR to certain GIS tasks, and the value of such implementation. Thus, although such work is outside the scope of this chapter, it is suggested that the above issues be challenged at: The user interface level; the system architecture level; and, the data level. This chapter will present guidelines that should steer the final development of a true, 4D VRGIS.

16.3 VR AND GIS: ISSUES OF SYSTEM INTEGRATION.

As already noted, the tendency for the use of VR in GIS has largely been to implement VR as an add-on to conventional GIS in order to visualise and distribute data views, often via the World Wide Web as Virtual Reality Modelling Language (VRML); in many cases the tendency is to use VR for indirect visualisation. Research is however now beginning to undertake the task of studying the implications of the use of VR in GIS, (Neves *et al.*, 1997). This, and similar research, must address the issues of integration at the following levels discussed below.

16.3.1 The user interface level

At the user level, the most important question to be considered is 'what are the best metaphors for interaction in a VRGIS?' By expanding this question further, we are presented with a set of research issues that will examine what is inherently the best paradigm by which users will be able to interact with a VRGIS, while maximising their cognition of the data or problem in hand.

How to best display, and make the user aware of location, attitude and time in a geospatial virtual universe? Will users require a Head Up Display (HUD), similar to those used in an aircraft, or a bank of virtual instruments and dials that display location, bearing, attitude, and altitude, to enhance their perception of their immediate location within the virtual space, and aid navigation?

Is a single (2D or 3D) or dual display (2D and 3D) metaphor required for particular tasks? Will having a real time linked 2D map indicating user location in the virtual space decrease the possibility of users becoming lost within the virtual space, as they will have a familiar display to reference their virtual location against? Thus, is a dual display metaphor always necessary?

What input and output devices traditionally associated with each genre are suited to use in VRGIS, e.g. is a digitising tablet and puck a suitable tool for interacting in a geospatial virtual universe? Is it suitable to carry out traditional GIS tasks such as digitising in 3D, or is interaction via 3D paradigms a barrier to efficient interaction? However, would a particular group of GIS users rather carry out tasks via an unconventional 3D methodology?

What is the exact value of VR in improving user cognition in GIS applications? A vast amount of research in the field of conventional VR has indicated that VR interaction technology affords greater levels of user cognition as the problem or data is placed into a more realistic context than conventional 2D. Is this the case in GIS, and in what instances? (see Human Interface Technology Lab (1998))

As noted, in conventional VR there has been very extensive research into these areas, however in such an application specific area as VRGIS, the research is severely lacking.

16.3.2 The system architecture level

At a system level the questions become much more fundamental to future VR and GIS integration. The most important question to be considered at the system level is how will the idiosyncrasies of existing GI systems, and existing VR systems or 3D engines, affect the intended functionality of a VRGIS? To illustrate this, a case study of a typical GIS function versus a typical item of VR functionality will be examined, that of Yon Clipping versus Zoom Extents.

16.3.2.1 Example: Yon Clipping vs. Zoom Extents

Although a technical term, Yon Clipping (YC) is in fact a simple concept. The word yon originates from yonder, the old English for over there (in the distance). The Yon Clipping plane is a function of most VR systems, and is the point beyond which objects in a simulation will not be rendered in the display, (regardless of whether they would be visible or not in real life). Zoom extents is referred to here as the typical GIS function of zooming out in the display to see an entire map or data set.

When constructing a virtual universe, it is necessary to specify the YC distance, which will be defined in the specified units of the universe. The purpose of the YC plane is to decrease the processing power required of the host computer. The virtual universe being rendered may contain a very large amount of polygons that make up the objects in the universe. For example, a terrain constructed from a common TIN or Indexed Face Set, will possibly contain 10,000 polygons, or realistically many more. The graphics card controlling the display and rendering process may however only be capable of processing and displaying 8,000 polygons a second at a realistic, and interactive frame rate, for example 24fps (frames per second). Thus, if a YC plane is specified that is within the spatial extent of the terrain, not all the terrain will be visible, as the section of terrain beyond the YC plane will not be rendered. Thus the number of polygons to be displayed will be less than the 8,000 per second limit of the card, hence the frame rate of the display will not decrease, but not all the terrain will be visible.

Therefore, using this technique, large terrains can be rendered, (yet not in their entirety), and an interactive frame rate can be maintained. It must be noted however that if the graphics card being used is very powerful and can render very large numbers of polygons, then a very large YC plane can be set that is outside the limits of the terrain, and hence the entire surface, and any other objects on it will be continually visible.

Thus far, the additional load of texture draping has not been mentioned; terrains and most other objects in a virtual universe will usually have an associated texture to increase its realism or meaning and this adds to the processing power required to display large or many objects. As the author suggests that VRGIS should not be exclusively the domain of high end graphics supercomputers and workstations, but of much lower to middle range desktop computers, use of techniques such as Yon Clipping will still be necessary to enable an efficient and interactive display.

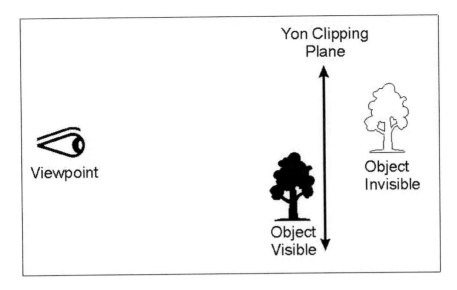

Figure 16.1 Yon Clipping: Objects between the viewpoint and the Yon Clipping plane will be visible, objects beyond the yon clipping plan will be invisible

Conversely, if incorrectly set, the YC plane may adversely affect graphics performance. If a terrain containing 10,000 polygons is to be displayed using a graphics card that is capable of displaying only 8,000 polygons a second, and a YC plane is specified that is outside the limits of the terrain, then all the terrain will be rendered or visible in the display, from all points of view. This however increases the load on the graphics hardware as it will have to render all polygons that are now visible. As the number of polygons to be displayed exceeds the capability of the graphics processor, a marked decrease in display frame rate will be encountered, possibly down to as low as 2 to 4fps. Low frame rates make navigation through virtual spaces difficult, as the display will no longer be linked in real time to mouse (or other user input device) movement, hence the value of interacting with the data through VR will decrease. Furthermore, if the number of polygons to be displayed becomes larger, the display may stop completely.

Bringing GIS into the equation generates further issues. For example, a possible likely use of VRGIS would be to examine view or horizon characteristics in application areas such as wind farm planning. The user may wish to examine the view from a certain viewpoint on the terrain to see if a set of wind turbines would be visible. The YC plane, if set within the spatial extent of the terrain, may mean that the terrain will cut off at the YC plane, hence the horizon visible from the viewpoint will be incomplete. Similarly if the user chooses to 'zoom extents' to a point beyond the YC plane, then the terrain will disappear altogether.

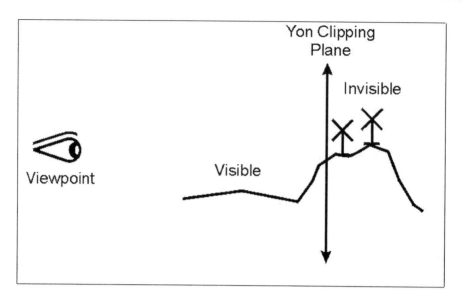

Figure 16.2   The effect of the Yon Clipping plane on horizon characteristics. Here the Yon Clipping plane intersects a terrain upon which a wind farm is situated. The portion of the terrain and wind farm beyond the YC plane will not be rendered; hence the horizon displayed to the viewer will be incorrect

Established standard computer graphics solutions exist to help solve this problem. These include back face culling of polygons, object switching, and intelligent data structures such as multi-resolution TINs (Muchaxo, 1997). However these techniques can cause problems for GIS applications due to the generalisation (in some or most cases) of the display while implementing these methods. For example, when visualising a terrain, if the horizon is too generalised as a result of the above techniques, the view may be incorrect, or outside acceptable levels of generalisation, thus presenting the user with a possibly unacceptable view of reality.

16.3.3   The data level

Finally, another fundamental issue is that of data input to a VRGIS system. The vast majority of GIS data is currently unsuited to conventional VR systems because a large amount of GIS data is two-dimensional. Furthermore, both VR and GIS suffer from a glut of often incompatible file formats within their respective genres. Many applications will import and export to the most common formats of each technology: 3ds, obj, nff, wrl, dxf, prj and slp and so on, (VR); and GIS: shp, dem, dxf, mif and ntf, etc.. One or two of these formats can be read by VR and GIS systems alike (e.g. dxf and dem) but there are often subtle differences in what is contained within the file generated in each respective genre. For example dxf files generated in a GIS may contain Z values, but the geometry often does not contain vertical and oblique faces, thus if imported into a VR/3D modelling application, the modes will lack detail and accuracy as a building might have a roof and footprint, but will have no walls due to the lack of faces. Common formats and standards are already a consuming issue of current GIS research and policy, and hence shall not be discussed here, however it is clear that methodical research will be required at this level.

The above examples are by no means exhaustive, but serve to illustrate the fact that if a true VRGIS is to be developed, certain aspects of each technology may need to be discounted. Other computer graphics (CG) techniques such as texture interpolation or anti-aliasing present further challenges. These techniques ensure that textures draped over objects are displayed in a smooth, un-pixelated fashion. These techniques however, although excellent for maintaining realism in computer games have implications for VRGIS application. As you move closer to an object, the anti-aliasing will cause the texture to become very blurred, often with a particular loss of edge definition. Will this affect our ability to read a detailed map that is draped over a virtual terrain? Careful consideration will have to be paid to how the functions of VR engines and CG techniques will affect their use in a VRGIS. Will current VR methods require extensive re-engineering? The answer is likely to be yes, as these challenges are further examined it is inevitable that further issues of incompatibility between VR functions and GIS functionality and application areas will come to light. The aim is to arrive at a development paradigm for VRGIS that is devoid of irrelevant or conflicting functionality, and serves to meet correctly the needs of specific tasks. This is an area that will hence require extensive research.

16.4 DEFINING AND JUSTIFYING 4D VRGIS: EXTENDING 3D GIS

3D GIS, the forerunner to VRGIS, could be classified as having five main functional components: Acquiring, structuring, storing and managing, processing and presenting spatial data (Pilouk, 1996). These are supported in turn by a variety of technologies, for example data acquisition can be supported by map digitising, photogrammetry, GPS, borehole and seismic etc. It is suggested here that true VRGIS would include a sixth functional component, that of 'Interaction'.

The output of conventional VR systems allow for an extremely high level of user interaction much more so than a conventional two or three-dimensional GIS. VR systems extend user interaction by supporting a variety of input and output devices. Visualisations can be rotated, flipped and zoomed in and out using both two and three-dimensional input devices. For example it is possible to examine and navigate a 4D virtual space using a 2D mouse, or on the other hand a 'space glove' that operates in three dimensions. Similarly display can take the form of several alternative types and dimensions. VR can be viewed on a flat desktop monitor in two and a half dimensions, or immersively using a Head Mounted Display (HMD) in true three dimensions. HMDs incorporate or are attached to trackers, such as the Pholemus device. Trackers sense head movement and hence adjust the display accordingly, giving the sensation of 'looking around' and environment or visualisation. The physical display of HMD's can be simple 2D displays, however certain devices also include a stereo option, so as the user will be immersed in a truly three-dimensional world.

A true VRGIS system will however have to include I/O devices traditionally associated with conventional GIS, such as digitising tablets and pucks. Research has already gone some way to incorporating the two paradigms, by integrating digitising tablets/pens, mice, orientation sensors and HMDs, the 'Virtual GIS Room' has been developed (Neves *et al.*, 1997). By incorporating the above devices, users are able to fully utilise the three cognitive spaces required for the efficient analysis and handling of spatio temporal information, haptic, pictoral and transperceptual, thus maximising the value of interaction with data.

One of the most crucial simplifications of conventional GIS and cartography, is its history of organising and displaying data in a way that is so far removed from reality, that of the 2D layer. On of the problems of this approach is that the level of abstraction from reality tends to be very high. This does have its value in that high levels of abstraction often help the naïve user to understand the information or problem being presented, as it is highly simplified. However it could be argued that this method of generalisation actually so far abstracts from reality that it can be hard to insert the information presented into a real world context. Secondly very little real world data only inhabits a 2D space, even if a phenomena or object truly lies on a flat plain, it is affected by time, (x, y, z, time), more accurate representation of reality however is x, y, z, t, (+ process, p, as a function of time). It is suggested here that a true VRGIS would be four or rather multi-dimensional; as to accurately model the real world or even abstract data, we must consider location (x, y, z), time and process. This is explored in the following section.

This paradigm will be required for several reasons. As mentioned above, 2D GIS can abstract data to a level where its context can be far removed. By rendering abstractions of data in three and four dimensions, the level of abstraction in most cases is placed into a more realistic context. Contours on a flat map can be hard for some people to interpret. By extending the display of contour data into three dimensions, the data is put into a more realistic context as the contours take on a more realistic, though still generalised, representation of reality; thus 'dense circles of orange lines' start to look like hills. By further extending the display into a fourth dimension process can be introduced. As VR displays are dynamic, rotational slides could take place in real, or accelerated time, using (real time or pre-processed) animation and morphing techniques, hence rather than examine and interpret before and after displays of phenomena in two or three dimensions, the event is put into context by the use of the fourth dimension. Thus the use of four dimensions in displaying spatial objects and phenomena would enhance the value and level of user interaction as abstracted information can be displayed in a more realistic context then 2D. Context can also be enhanced by the use of 3D sound, and through the use of additional hyperlinked 2D imagery and 2/3D geometries, or in other words contextual qualitative information. It is also suggested in section 16.3 of this chapter that, a dual metaphor interface should be an option of VRGIS. Using a dual window interface e.g. 2D mapped representations can be displayed next to a real time linked 3/4D rendering, thus context of each display can be enhanced through direct comparison to the other.

Further to this, many GIS application areas have to incorporate time as being fundamental to their management and modelling process, for example fisheries management, (Lee and Kemp, 1998). Similarly, time extends the views available of spatial information: data can be explored at varying scales, levels of detail and additionally at various temporal granularities, by the minute, hour, day, week and so on.

To arrive at a definition of VRGIS we must therefore consider a system that utilises the common aspects of both technologies and the more application specific functionality of each genre. Such a system must therefore have the capability of:

(a)   Creating direct and indirect representations of determinate and indeterminate spatial entities, at varying levels of abstraction, in two, three, and four dimensions, (4D + Geo/Spatio Temporal attributes)
(b)   Supporting process, time, behaviour and evolution.
(c)   Visualising / rendering the abstracted data in two, three, four/multi-dimensions. (Both in virtual or real world co ordinates / time).
(d)   Facilitating both high and low level interaction with the visualisation/data:
      High Level: Immersive or desktop navigation of the Virtual Universe.
      Point and click query/hyperlinking to multimedia data and virtual (real world co-ordinates) space/time/process data.
      Low Level: Allow the performing of spatial queries and operations upon the visualisation itself, as well as the spatial data source.
(e)   Capturing, updating and storing spatial data.
(f)   Facilitating spatial analysis functions in real time, on the visualisation, such as the calculation of volume, surface area, centre of mass, nearest neighbour etc.
(g)   Storing and organising data in as near a real world paradigm as possible, thus a multi-dimensional scene graph.
(h)   Mirroring multi-dimensional dynamic interaction through two, three, and four-dimensional interfaces in real time, i.e. what happens in a map view dynamically happens in the virtual view and vice versa.
(i)   Supporting input and output devices native to both GIS and VR, for example, scanners (2 and 3D), digitising tablets/pucks, plotters, data/cyber gloves, HMD (Head Mounted Displays), space-balls, motion trackers/attitude sensors and conventional monitors.
(j)   Supporting and generate LOD (Level of Detail) in both geometry and textures, which are constructed to suit the users level of geographic knowledge, or the intended end use.
(k)   Allowing the maintenance of existing models, including the merging of data sets to generate new models.
(l)   Reading and exporting a limited set of separate VR and GIS file formats, and generating a native VRGIS format.
(m) Allowing the developer/user to set lighting and viewpoint attributes, as well as predetermined paths through the simulation.

The above points are not intended as the complete functionality set of a VRGIS environment or system, but are a theoretical outline of the concepts that should be considered as necessary in the development of the true VRGIS paradigm. It is accepted that conventional two and three-dimensional GIS already contain some of this functionality, but a truly integrated system that contains the complete set is yet to evolve. Thus, taking these points into consideration, a 'true VRGIS' could be defined as:

'A multi-dimensional, computer-based environment for the storage, modelling, analysis, interrogation of, and interaction with, geo-referenced spatio temporal data and processes.'

16.5 VRGIS CONCEPTUAL ARCHITECTURE: SPACE, TIME, COORDINATE SYSTEMS, GIS FUNCTIONS AND THE SCENE GRAPH.

In this section, the theoretical design concepts of a 4D VRGIS architecture are discussed. Rather than concentrate on technical implementation, an examination of how a VRGIS might deal with space, time, co-ordinates, measurement, GIS functionality and universe structure is proposed.

16.5.1 Space and co-ordinate systems

The real world may be viewed as a 'space', a collection of spatial objects and the relationships between them (Gatrell, 1991). Each spatial object within the space occupies a subspace to define its own spatial extent, defined by a set of spatial co-ordinates and properties that describe the location, (Smith *et al.,* 1987). Different sets of relations will define different types of space, for example metric space is based on distance relationships.

Real space is an unbounded region consisting of numerous objects and relationships. *A space*, S (finite set), is a view of reality in which the context is defined for describing the aspects of reality relevant to a particular discipline. In this case S is a collection of objects (O) and their relationships (R), denoted S={[O],[R]}, (Pilouk, 1996).

How to represent this view of reality, S, depends on the demands of the particular discipline. As VR and GIS describe space, objects and their relationships in different ways, an examination of the two methodologies is required in order to understand the implications of each for VRGIS.

Most conventional 3D and VR systems describe space (S) using a scene graph. As mentioned the scene graph is a hierarchical arrangement of nodes that represent objects, their attributes, and position in three or four dimensions. To fully understand the scene graph, both its organisational structure, and secondly, the co-ordinate systems must be described. In conventional computer graphics there are four major co-ordinate systems: model, world, view and display. The world system, the most important co-ordinate system in a graphical simulation, or universe system is a fixed, object independent space in which entities are positioned. Objects imported must translate their arbitrary co-ordinate system into the world co-ordinate system, through the scaling, rotating and translating of their geometry. Secondly, it is in the world system that the position and orientation of cameras and lights are specified.

VRML for example uses the above co-ordinate system, often referred to as the right hand rule for 3D axes (Figure 16.3). Although this system is fairly generic, it is most certainly not the rule for all VR and 3D systems. Certain packages, for example, us a system in which the normally positive values become negative, i.e. a reverse co-ordinate system.

Existing 3D GIS uses a co-ordinate system where the plane upon which data is mapped and displayed is the X, Y plane, with elevation in Z. Thus in order to import a geospatial object into a virtual universe, co-ordinate transforms will have to take place in order to maintain the spatial integrity of the imported object. As mentioned VRML for example, uses the right hand rule co-ordinate system, with the default viewpoint looking along the axis from the origin. A 3D vector rotation axis (x, y, z magnitudes) and rotation angle defines viewpoint orientation. As this has implications for transformation and extraction of real world co-ordinates and view orientation, proposals for standard co-ordinate structure have already been recommended (Moore *et al.,* 1997). Such a standard will, with regard to indexed face set/TIN models, provide consistency with equivalent grid models, and reduce the magnitude of numbers stored in the model (Moore *et al.,* 1997). Further to this a geoVRML working group has been established to further develop and research a certain level of geospatial functionality into future VRML releases (Iverson, 1998). It is recommended here that again these issues be examined further, but as discussed earlier, research must extend beyond issues of geo-referencing, time referencing, terrain representation, level of detail, accuracy and data exchange, and examine the issues of integrating actual GIS tasks into the Virtual Reality.

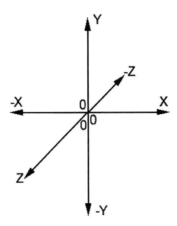

Figure 16.3 The 'Right Hand Rule' of the VRML co-ordinate system

16.5.2 Space and measurement

The most common unit of measurement in VR co-ordinate systems is the metre, although most packages allow the user to work in, or translate to other unit measurements such as inches or miles. Using single or double precision floating point representations of three-dimensional co-ordinates is sufficient to accurately display content rich 3D simulations. As already noted, conventional VR systems use an origin of 0,0,0; as one extends to 100 miles from the origin, representable points become quite quantitised, to approximately a third of an inch or coarser still. Likewise, when working in extremely small co-ordinates, the same errors occur, (Sun Microsystems, 1997). To combat this, The J-3D API Version 1 1997 proposes a high-resolution addressing/co-ordinate system, by implementing 256-bit positional components.

The J-3D API specifies that each XYZ co-ordinate be defined with a 256-bit fixed-point number, the fixed point is at bit 128 and the value 1.0 is defined by exactly one meter. Using such a technique it will be possible to define a virtual universe that is in excess of several hundred billion light years across, and include in it objects that are less than a proton in size (below Planck length). Further to this, a 256 bit can represent almost any single-precision floating-point value exactly. Thus quantitisation of location at long distances from the 0,0,0 origin of a conventional VR universe will be removed allowing for the exact positioning of spatial objects. It is suggested that a VRGIS would require a similar method of co-ordinate representation as this would remove the generalisation of co-ordinates at large distances from the universe origin, and allow the modelling of real world sized simulations and spaces.

16.5.3 Scene graphs, time and GIS functionality

A scene graph is a hierarchical arrangement of nodes (e.g. geometry, light, sensors, time, anchors, sound, groups, transform, or geo-location and so on) representing objects in a simulation. Scene graphs are constructed from a collection of nodes or objects that are connected to other objects, and describe the contents of a virtual universe. A detailed comparison of different API and ISO Specification scene graphs will not be conducted here, but a summary of the major component or node types offered.

Group nodes are the glue of a scene graph; they group together child nodes into universe subsections. Child nodes are the universe 'contents', i.e. buildings, people, backgrounds, sounds and also behaviours associated with objects, finally lights and viewpoints are included as child nodes. Sensor nodes establish when a condition is true based on time, actions, or distance changes in the virtual world. Commonly scene graphs are constructed in the appearance of an inverted tree and contain a simple description of the entire VR universe or scene. This will include geometric data, attributes and viewpoint information. Scene graphs may also contain script nodes, which add extra levels of functionality into the simulation. Script nodes can be used to define custom nodes that extend the scene graph beyond its basic component set. For example, a script node might contain instructions that define the specific combination of virtual key presses that would have to be carried out in order to open a virtual door.

It is suggested that a VRGIS would be based around the concept of a scene graph. The scene graph concept is particularly suited to the description of geo/spatio-temporal data for several reasons: Spatial objects can be described in four dimensions, free from the layered approach. Objects can belong to groups, (or be described as individual, independent), components that are governed by the actions/attributes of their parent. Objects can move through the hierarchy of the scene graph and transport/or adjust their properties/attributes according to their relocation. Objects can be acted upon by process/time, or specified with properties that adjust to, or are governed by their spatial location, or interaction with process 'zones' in the virtual universe.

Time would be represented and described by a clock object, with two subtypes. The abstract clock object would provide the notion of time supported by the environment. Time would function on the existence of two non object types: Time, representing elapsed 'ticks', and TimeUnit, representing Tick duration/unit. The first sub type of the clock object would be the SystemClock, providing real time information, i.e. the real system time. A start time would have to be provided, to thus give the SystemClock object a value. The second sub type of clock would be the Timer object, that would model the basic functionality of a stopwatch. Thus the Timer object is a finite state machine, with state transitions as paused, running and stopped. (PREMO, 1997).

It is suggested that GIS functionality should be integrated into the scene graph, as a node similar to that of the script node, hence a GIS Function node. By placing the GIS functionality into the scene graph, GIS tasks would be carried out 'within' the data, hence placing task at the system/data level, in a near real world, multi-dimensional context. Furthermore assigning objects within the scene graph with attributes of space/time, such as process and behaviour, the multi-dimensionality of the simulation is further expanded. Thus geo-referenced spatial entities can be simulated and interacted with in a real time, dynamic, evolutionary context.

16.6 CONCLUSION

This chapter has served to illustrate some of the challenges and opportunities faced by the future integration of VR and GIS to arrive at true, 4D VRGIS. It is intended that the challenges presented here are fully investigated, and case studies produced to eliminate the traps and dead ends that future development may encounter. A further examination is also required, perhaps more importantly, of what is meant in the case of VRGIS by the term 'Virtual'. Are we aiming to exactly recreate reality, or still present only levels of abstraction to the user? If we are looking to recreate reality, then the issues of real time data acquisition and processing must be considered, as how can a model be said to represent reality if it is not real time? Finally, it must be noted that VRGIS should not be seen as a replacement for conventional GIS, but as a tool that will enhance and extend conventional GIS operation into multiple dimensions, affording greater levels of user interaction and cognition.

REFERENCES

Câmara, A., 1996, Virtual Reality and Visualization of Water Resources. In International *Symposium on Water Resources Management*, (Aachen).

ESRI, 1991, *Understanding GIS, The ARC/INFO Method*, Version 7, (Redlands: California).

Gatrell, A.C, 1991, Concepts of space and geographical data. In *Geographical Information Systems: Principles*, Vol. 1, Goodchild, M. and Rhind, D. (eds) (Longman).

Haklay, M., 1998, A Survey of current trends in incorporating virtual reality and geographic information systems. In *Proceedings GISRUK '98*, (Edinburgh).

Human Interface Technology Lab, 1998, Human Interface Technology Lab, Washington, Bibliography, http://www.hitl.washington.edu/publications/index.html.

Iverson, L., 1998, GeoVRML Working Group, http://www.ai.sri.com/geovrml.

Kalawsky, R.S.,1993 *The Science of Virtual Reality and Virtual Environments*. (Addison-Wesley).

Lee, H. and Kemp, Z., 1998, Supporting complex spatiotemporal analysis in GIS. In *Innovations in GIS 5*, Carver, S. (ed) (London: Taylor and Francis).

Muchaxo, J., 1997, Multi-Scale Representation of Large Territories. In *Spatial Multimedia and Virtual Reality*, Câmara, A., and Raper, J. (eds) (London: Taylor and Francis).

Neves, N, Silva, J.P., Gonçalves, P., Muchaxo, J., Silva, J. and Câmara, A., 1997, Cognitive Spaces and Metaphors: A solution for Interacting with Spatial Data. *Computers & Geosciences*, Vol. 23(4), pp.483-488.

Neves, J.N., Gonçalves, P., Muchaxo, J., Jordão, L., Silva, J.P., 1997, Virtual GIS Room. In *Spatial Multimedia and Virtual Reality*, Câmara, A. and Raper, J. (eds) (London: Taylor and Francis).

Pilouk, M., 1996, *Integrated modelling for 3D GIS*. ITC Pub. no. 40, (Netherlands: ITC).

PREMO, 1997, *PREMO Draft International Standard*, (ISO/IEC 14478-1).

Raper, J., McCarthy, T., and Williams, N., 1997, Integrating ArcView and Geographically Referenced VRML Models in Real Time. In *Proceedings of the ESRI User Conference 1997*, (San Diego).

Smith, T.R., Menon, S., Star, J.L., and Estes, J.E., 1987, Requirements and principles for the implementation and construction of large scale geographical information systems, *International Journal of Geographic Information Systems*, Vol. 1, pp.13-32.

Sun Microsystems, 1997, *Java 3D API Specification*, (Sun Microsystems).

PART V

Informing Through Modelling

17

An integrated temporal GIS model for traffic systems

Adam Etches, Christophe Claramunt, Andre Bargiela and Iisakki Kosonen

17.1 INTRODUCTION

Currently one of the most important challenges for GIS is to generate a corporate resource whose full potential will be achieved by making it accessible to a large set of end-users. For instance, a GIS managing both static urban data and dynamic traffic information could provide an integrated geographical reference to the management of traffic flows leading to the improvement of the quality of transport systems. However, current GIS software and interfaces do not encompass the set of functions to make this technology compatible with simulation models used for traffic control and for traffic planning. The integration of GIS and simulated traffic systems is likely to be a challenging and worthwhile objective for both user communities, whose needs are not satisfied by loosely connected set of existing solutions. In particular, the current form of integration of a GIS and simulation models still use loose coupling methods, such as passive file transfers and separate user interfaces. This poor level of integration is often a result of the different model paradigms used within GIS and modelling systems, and the fact that any integration solution implies a re-design and re-implementation of existing software.

This chapter proposes the design and development of a temporal GIS (TGIS) for simulated traffic systems that permits the manipulation of a common reference database. The project aims to investigate and identify a database model that will favour an active collaboration, in terms of data integration and tasks sharing, between simulation models used in the context of traffic system and a TGIS. The projects main areas will be centred on key features that are involved in the design of a traffic system such as geographical data characteristics and behaviour. The chapter is organised as follows: Section 17.2 presents the basic concepts involved in the simulation of traffic systems; section 17.3 introduces the related database modelling issues; section 17.4 develops the principles of our traffic data model; section 17.5 describes the resulting model with a conceptual and graphical representation; an object-oriented description of the model is proposed in section 17.6; and finally, section 17.7 draws conclusions.

17.2 SIMULATION OF TRAFFIC SYSTEMS

The increasing pressure to make the best sustainable use of the capacity or urban road networks has led to the development of sophisticated traffic management and control strategies which take into account real-time measurements of traffic flows, the interaction of various modes of transport and the priorities of private, public transport and emergency vehicles. The evaluation of the efficiency of these strategies relies on the availability of suitable simulated environments, within which it is possible to re-play pseudo-random traffic occurrences for the purpose of a comparative study. In theory, any phenomena that can be reduced to mathematical data and equations can be simulated. In practice, however, simulation is a difficult task because most natural phenomena are subject to an almost infinite of influences.

Particularly the simulation of traffic needs to reflect a lot of subjective decisions made by drivers which concern their individual driving styles, route selection and so on, about which there is usually little or no measured information. In this context, it is suggested that traffic simulations should always be considered at several levels of abstraction; from microscopic, with well defined, deterministic vehicles, to macroscopic, with probabilistic traffic descriptions. We define a traffic system as a real-world environment that characterises the behaviour of vehicles within an urban network. A simulated traffic system is a computing resource that allows the real-time prediction, control and monitoring of urban traffic flows. Our example system is currently oriented toward the representation of a simulated traffic system. A simulator must model each vehicle in the network as a separate entity (i.e. microscopic level) or as traffic as an aggregated set of entities (i.e. macroscopic level).

The proposed TGIS provides a means of supporting a full spectrum of traffic simulations which need to have a real-time access to various levels of aggregation of traffic data and to execute concurrently to provide a realistic description of urban traffic. The simulation model represented in our project integrates these complementary abstraction levels: the macroscopic level with the Probabilistic Adaptive Simulation Model (Peytchev, 1996) and the microscopic level with the HUTSIM micro-simulation system (Kosonen, 1995). Both models integrate probabilistic characteristics or dynamic navigation of car or traffic flows moving through the urban region (e.g. the description of a car behaviour as probabilistic rather than the deterministic route assignment to a car). Represented data include dynamic features such as detected or estimated traffic flows, the management of events such as traffic accidents and incidents and the analysis and estimation of distributed processes in discrete locations (Kosonen, 1995; Peytchev, 1996). Approaches to microscopic models involve the representation and/or optimisation of complementary data such as driver behaviour, vehicle characteristics and performance, road components, lanes structure and geometry, and their relationships. The simulation tries to minimise the amount of data required to give reasonably accurate simulation results. The HUTSIM micro-simulation system provides a model which is flexible and relatively detailed and potentially adapted to different traffic system configurations (Kosonen, 1995).

17.3 DATABASE REPRESENTATION ISSUES

Database integration issues include the development of general methods and techniques facilitating the use of geographical data in the context of time-varying information, and the integration of traffic data as a new component of GIS. The fact that the next generation of information system will be based on co-operative information systems is now widely accepted (Papazoglou, 1992). Several proposals have been defined for integrating GIS with real-world models. They are based on the definition of external workflow, i.e., a sequence of programs (Alonso, 1997), and on the identification of an integrated database schema and translation functions (Abel, 1997). However these proposals are generally oriented toward the representation of environmental applications and are therefore not adapted to the specific dynamic characteristics of traffic systems.

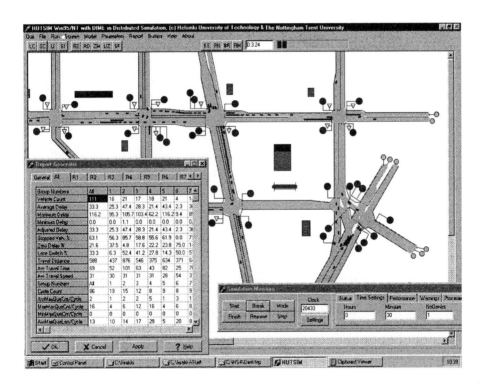

Figure 17.1 HUTSIM micro-simulation system

From a methodological point of view, a successful integration of a GIS with a traffic system implies the definition of different, and successive, collaborative pathways at the database, processing and interface levels. We propose a database integration approach, in other words a semantic and logical data integration, that identifies a set of minimal modelling features that constitute a set of pathways between a GIS and a traffic system. A conceptual design of geographical and traffic data is accomplished to support the development of an interoperable database that includes a set of functional modelling primitives. Such a framework provides a step toward the development of software integration at both the processing and interface levels. The additional temporal component of traffic systems may be considered and integrated as an additional dimension to the GIS structure.

Temporal GIS has received an increasing amount of recognition over recent years (Al Taha, 1994). TGIS proposals benefit from advances in temporal database research. The temporal database approach provides facilities for storing past, current and future data (Tansel *et al.*, 1993; Snodgrass *et al.*, 1995). Many conceptual spatial data models have been defined in order to integrate the time dimension within spatial data models (e.g. Tryfona (1995), Story (1995), Bédard (1996), De Oliveira (1997)). Nonetheless, these spatiotemporal models do not provide a complete support for the database modelling of physical processes involved in the description and simulation of real-world phenomena.

Building an efficient and consistent system to operate simultaneously on absolute and relative views of space and time implies the integration of geographical, temporal and thematic components (Peuquet, 1994). This must be addressed at the event and process levels to discover how changes happen and how entities are related into spatio-temporal interaction networks (Claramunt and Thériault, 1995, 1996). Processes are of a particular interest in a representation and study of environmental and urban applications. They allow the understanding of the dimensions of real-world systems. Our modelling approach integrates the time component with current spatio-temporal design methods as it allows the representation of real-world processes within the database system. The method is based on an object-relationship method extended to the modelling of dynamic geographical applications.

The spatial database method we use to support the description of the traffic system is based on the MADS object-relationship model (Parent *et al.*, 1997). MADS is developed from the Object Database Standard (Catell, 1994) and aims at the same objective as the OpenGIS specification proposed as a reference for geographical data models (OGIS, 1996). It can be adapted to any object-oriented or entity-relationship spatio-temporal model. This object-relationship model identifies a set of principles for the representation of entities and relationships. It has been extended to the representation of spatio-temporal processes (Claramunt *et al.*, 1997).

MADS represents space at different levels: object type, attribute and relationship type. Spatial types are represented in schema diagrams by specific icons. Spatial relationships may be defined among spatial objects to express integrity constraints on their geometry's or the application spatial semantics. Any object type or relationship type can have spatial attributes whose domain is one of the abstract spatial types. Each object is referenced by a unique and time-invariant system defined identifier. Defining an object type as temporal implies that the database will represent all the different object instances/values, associated with their corresponding valid times. Accordingly and to give a high flexibility to the database design task, processes are either described as object types, relationships type or as attributes. Processes share a set of non-temporal and temporal attributes. The model supports both object and attribute versioning allowing a flexible representation of temporal properties (respectively grouped and ungrouped temporal models according to Clifford (1995)).

17.4 TRAFFIC SYSTEM MODEL PRINCIPLES

In order to integrate traffic system components and complementary simulation levels, our spatio-temporal model represents data in space and time at different abstraction levels. For instance a road may be represented at both logical and geometrical levels. A road can be described as an aggregation of interconnecting lines and nodes from a logical view, but the same road can be considered to have a form and a surface from a geometrical view. The database model integrates the following dimensions: (1) the description level, i.e., objects and properties described as object attributes; (2) the geographical level that includes logical and geographical representations in space and (3) the temporal level (objects life, object property changes, processes).

From a database point of view, the traffic system domain is composed by static and dynamic characteristics. Static components are data objects such as street layouts and topological features such as roads. These features are considered as static within a traffic system that examines short time scales in object transformations. Dynamic components (measured or simulated) are objects that change on a relatively short time scale from an application point of view (e.g. cars, traffic queues, traffic signals). Our objective is to identify significant properties about the dynamic traffic mechanisms, to explicitly record relationships among entities involved in traffic processes, and to model causal relationships when they are identified from an application point of view. We consider different abstraction levels in both space and time in order to integrate different levels of spatial and temporal details. At each abstraction level, computing models generate a set of database primitives from actual and simulated measurements (Figure 17.2).

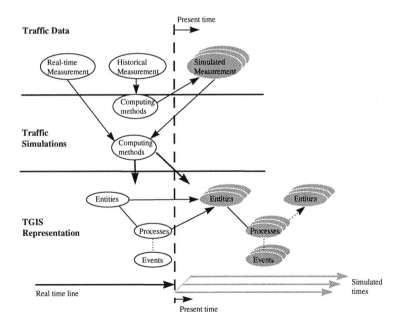

Figure 17.2 TGIS traffic model principles

The modelling primitives used within the TGIS model are as follows:

- *Entities*. An entity represents a real or interpreted concept (e.g. a network, a vehicle). Aggregated and simulated entities may be defined at the microscopic (e.g. a car) or macroscopic simulation levels (e.g. traffic).
- *Processes*. A process is an action which modifies entities, the action is the interest (e.g. a vehicle displacement, a traffic generation).
- *Relationships*. They include thematic, spatial and temporal relationships. Spatial relationships may be defined among spatial objects to express integrity constraints on their geometry or the application spatial semantics (e.g. a signal unit is connected to a lane). Temporal relationships represent precedence properties between different entities (e.g. an accident precedes a traffic congestion). Causal relationships are particularly relevant in the case of traffic systems (e.g. a road accident causes a network perturbation).
- *Facts* and *properties*. They enable observation changes and allow the description of processes, entities and relationships.

These primitives are represented as objects, attributes or relationships within the database model, they provide a modelling support for the analysis of the traffic system behaviour. They have a given truth-value under a temporal period or instant. Time representations vary from small to large temporal granularities (e.g. second for a vehicle displacement, week for a roadwork) with immediate consequences on historical database strategies. The TGIS is used both as an agenda (e.g. measured real-time flows) and as a predictor (e.g. estimated traffic flows).

17.5 TRAFFIC SYSTEM MODEL

The database design is based on a complementary analysis of the HUTSIM microscopic traffic simulation system (Kosonen, 1995) and of the Probabilistic ADaptive SImulation Model PADSIM (Argile, 1996; Peytchev, 1996). Data properties are analysed at the microscopic (e.g. representation and simulation of vehicle displacements) and macroscopic abstraction levels (e.g. representation of aggregated traffic displacements). For the purposes of clarity the database schema of the simulated traffic system is presented as three overlapping sub-schemas which describe geographical data, traffic data and simulated data, respectively. It is presented using a graphical language which models application objects, relationships and processes. This visual presentation is based on the MADS object-relationship model which describes the schema components and properties by specific icons and graphic symbols (Parent *et al.*, 1997; Claramunt *et al.*, 1997). We will now describe the main elements of this model.

The geographical data sub-schema describes the location reference of the network system (Figure 17.3). It does not include temporal properties as the network and its components, supporting the traffic system, are considered as a static system for the temporal life of the application.

From the higher abstraction level the geographical data model describes the application *Network* which is an aggregation of roads. A *Road* is defined by an aggregation of *Nodes* and *Road Segments*. A *Road Segment* is decomposed into *Lanes* which are represented by a single directional line spatial data type. The *Road Segment* includes two spatial data types, line and geocomposite, that represent the logical and geometrical views, respectively. The geocomposite spatial data type is derived from the *Lane* spatial data type. A *Lane* acts as key location reference for the management of traffic data and the application of traffic simulations.

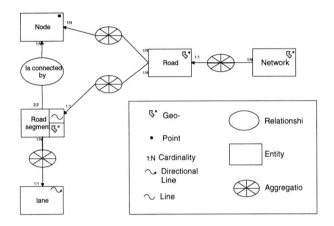

Figure 17.3 Geographical database schema

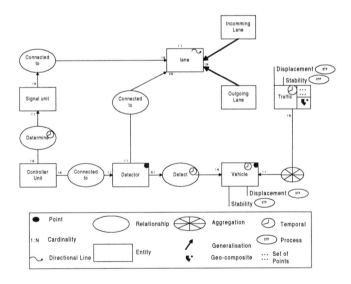

Figure 17.4 Traffic database schema

The traffic sub-schema represents data directly used and/or generated by the simulation traffic system (Figure 17. 4). It includes geographical and temporal objects as well as spatio-temporal processes that characterise object changes in space and time. The *Lane* object is specialised in *Incoming Lane* and *Outgoing Lane* objects to model the locations where *Vehicle* and *Traffic* instances are generated and removed, respectively, within the traffic simulated system. The relationship between the microscopic and macroscopic simulation levels is represented by an aggregation relationship that aggregates *Vehicles* into a *Traffic* object. Spatio-temporal processes include *Displacement* and *Stability* processes that are represented as attributes of both *Vehicle* and *Traffic* objects. The location of *Vehicle* and *Traffic* objects within the network is monitored by a *Detector* object. *Signal Unit* status is determined by the *Controller Unit*.

Finally, the simulation sub-schema includes the computing objects that generate vehicle and traffic objects at the microscopic and macroscopic simulation levels, respectively. *Vehicles*, represesented as objects, are specialised in *Virtual Vehicle* and *Real Vehicle* objects in order to represent the fact that vehicle instances can be generated by the simulation system or physically detected, respectively. The *Traffic* object includes two spatial data types, set of points and geocomposite, that represent the macroscopic and microscopic simulation levels, respectively. The geocomposite spatial data type is derived from the *Vehicle* spatial data type. The global database schema is presented in Figure 17.5.

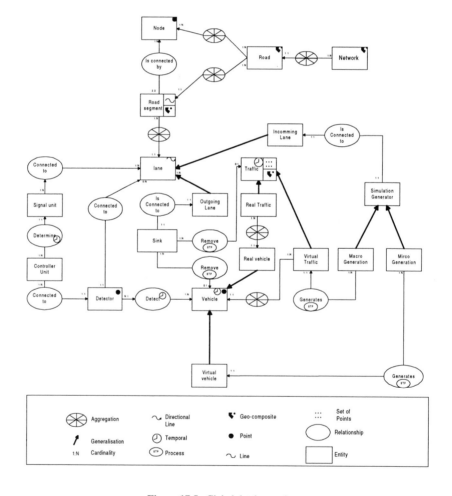

Figure 17.5 Global database schema

The global database schema provides a graphical view of the simulated traffic system database. Such a medium provides an efficient communication method between database designers and simulation experts. In fact the final schema is the result of an interactive design process that has implied discussions, exchanges and at least a common understanding of the application semantic. Specialised database concepts such as object-oriented logical schemas or computing program used within the traffic system are not so appropriate as their understanding is restricted to experts of the application domain.

17.6 OBJECT-ORIENTED SPECIFICATIONS

The conceptual database model is translated into an object-oriented representation. Next, we will present the textual descriptions that illustrate the main concepts used. Temporal granularities are represented at the object, attribute, geometrical, and spatio-temporal levels. For example the object *Vehicle* has a temporal granularity of a *Day* (Table 17.1). The geometry of the *Vehicle* object is recorded with a temporal granularity of a *Second*. The activity of *Vehicle* objects within the network system is described by the *Displacement* process, the absence of movement by the *Stability* 'process'. Both processes are specified by an *Interval* temporal type. The successive locations of a vehicle are recorded with a temporal granularity of a second (*Detected By*). *Vehicle*s are removed by a *Sink* object.

Table 17.1 Vehicle example

```
OBJECT VEHICLE
TEMPORAL DAY
GEOMETRY:              POINT TEMPORAL SECOND,
STP:                   (DISPLACEMENT INTERVAL TEMPORAL SECOND,
                       STABILITY INTERVAL TEMPORAL SECOND)
ATTRIBUTES:            (DETECTED_BY:  DETECTOR TEMPORAL
SECOND,
                       REMOVED_BY:    SINK TEMPORAL SECOND)
END VEHICLE
```

The generalisation abstraction allows the inheritance of temporal, descriptive and geometrical attributes, as well as spatio-temporal processes from the generalised object to the specialised objects. For example *Virtual Vehicle* and *Real Vehicle* objects both inherit the *Vehicle* properties (Table 17.2). A *Virtual Vehicle* removed time is equal to its disappearance time while the disappearance of a *Real Vehicle* within the simulated system does not imply his 'real' disappearance. Thus the system time of a *Real Vehicle*, i.e., the lifetime of a *Real Vehicle* within the network, is recorded by a time interval (*System Time*). A temporal attribute records the generation location and time of a *Virtual Vehicle* (*Generated By*). *Displacement* and *Stability* temporal values will be derived from locations detected within the network for *Real Vehicle*s whereas they will be directly defined by the simulation system for *Virtual Vehicle*s.

Aggregation relationships are represented by bi-directional links. The following example presents the *Lane* to *Road Segment* aggregation and its representation using the textual description (Table 17.3). The *Road Segment* specification includes two geometrical attributes that describe the logical geometry defined by a line type (*Global*

Geometry), and an aggregated geometry derived from the *Lane* objects (*Local Geometry*), respectively.

Table 17.2 Real versus Virtual Vehicle

```
OBJECT REAL_VEHICLE INHERIT VEHICLE
SYSTEM_TIME INTERVAL SECOND,
ATTRIBUTES:     ( VEHICLE_TYPE: STRING)
END REAL_VEHICLE
```

```
OBJECT VIRTUAL_VEHICLE INHERIT VEHICLE
ATTRIBUTES:     ( GENERATED_BY MICRO_GENERATOR SECOND)
END VIRTUAL_VEHICLE
```

Table 17.3 Road Segment - Lane

```
OBJECT ROAD_SEGMENT
  GLOBAL_GEOMETRY:   LINE,
  LOCAL_GEOMETRY:    GEOCOMPOSITE */ Derived from the Lane geometry,
  ATTRIBUTES:        ( FROM_NODE:              NODE,
                       TO_NODE:                NODE,
                       TYPE  :                 STRING,
                       SPEED_LIMIT:    INTEGER,
                       COMPOSED_BY:    SET(LANE) )
  END ROAD_SEGMENT
```

```
OBJECT LANE
  GEOMETRY:             DIRECTIONAL LINE,
  ATTRIBUTES:           ( LANE_TYPE:           STRING,
                          MEMBER_OF:           ROAD_SEGMENT)
  END LANE
```

The specification of the *Traffic* object illustrates the inheritance mechanism. *Real Traffic* and *Virtual Traffic* objects inherit their properties from the generalised *Traffic* object. The aggregation concept is represented whereby a *Real Traffic* is composed by a set of *Real Vehicles* and as a *Virtual Traffic* is composed by a set of *Real Vehicles* and *Virtual Vehicles*. A *Virtual Traffic* is generated by a *Macro Generator* whereas the existence of a *Real Traffic* is directly detected within the network system. The *Traffic* dynamism is represented by *Displacement* and *Stability* spatio-temporal processes (as for the *Vehicle* representation). The *Traffic* location within the network is temporally detected at a *Lane* location by a *Detector* object. A *Traffic* is removed from the simulation system by a *Sink* object which is located at an *Outgoing Lane* location. The *Traffic* object integrates two geometrical representations. These geometrical representations describe the logical level by a geometry defined as a set of points (*Macro Geometry*), and the local level by a geometry derived from those of the set of *Vehicles* involved in the aggregation link (*Micro Geometry*). As for *Real Vehicle*, the system time, i.e., the lifetime of a *Real Traffic* within the network is recorded by a temporal interval (*System Time*). Similarly a temporal attribute records the generation location and time of a *Virtual Traffic* (*Generated By*).

Table 17.4 Traffic versus Real Traffic and Virtual Traffic

```
OBJECT TRAFFIC
  TEMPORAL SECOND
  MACRO_GEOMETRY:      POINT,
  MICRO_GEOMETRY:      SET OF POINTS   */Derived from its set of vehicles,
  STP:                 (DISPLACEMENT INTERVAL TEMPORAL SECOND,
                       STABILITY INTERVAL TEMPORAL SECOND)
  ATTRIBUTES:          ( FLOW_RATE:              INTEGER
                       COMPOSED_BY:   SET(VEHICLE)
                       DETECTED_BY DETECTOR TEMPORAL SECOND,
                       REMOVED_BY SINK TEMPORAL SECOND )
  END TRAFFIC
```

```
OBJECT REAL_TRAFFIC INHERIT TRAFFIC
  SYSTEM_TIME INTERVAL SECOND
  END REAL_TRAFFIC
```

```
OBJECT VIRTUAL_TRAFFIC INHERIT TRAFFIC
  ATTRIBUTES:     (GENERATED_BY MACRO_GENERATOR SECOND)
  END VIRTUAL_TRAFFIC
```

17.7 CONCLUSION

This modelling approach permits the representation of database features involved in the design of a TGIS model for simulated traffic systems. The integrated model identifies a set of complementary modelling primitives (entities, spatio-temporal processes, relationships and properties) that together represent a database modelling reference. The resulting model integrates the geographical, temporal and simulated dimensions within an homogeneous database representation. The specific knowledge of the traffic system is represented by database design patterns such as the inheritance, aggregation and multi-representation concepts. Temporal properties are represented at both the life and motion levels and by an integration of spatio-temporal processes at the attribute and relationship levels. The model integrates both real objects detected within the network and virtual objects derived by the simulation model.

The database modelling approach provides a useful support for the identification of a traffic system behaviour at both the macro- and micro-simulation levels. It provides a graphical support for a collaborative integration of database designer and simulation engineer views as each domain own its proper set of concepts and semantic terms. A prototype implementation is under development. In order to realise a rapid prototype the object-oriented model is currently translated to a geo-relational GIS in a Windows environment. Further issues concern the development of data processing mechanisms with the HUTSIM simulation system and the development of an integrated interface.

REFERENCES

Abel, D. J., Taylor, K. and Kuo, D., 1997, Integrating modelling systems for environmental management information systems. *SIGMOD Record*, Vol. 26(1), pp. 5-10.

Al-Taha, K., Snodgrass, R. and Soo, M., 1994, Bibliography on spatio-temporal databases. *International Journal of GIS*, Vol. 8(1): 195-203.

Alonso, G. and Hagen, C., 1997, Geo-Opera: Workflow concepts for spatial processes. In *Proceedings of the 5th International Symposium on Spatial Databases SSD'97* (Berlin: Springer-Verlag), LNCS 1262, pp. 238-258.

Argile A., Peytchev, E., Bargiela, A. and Kosonen, I., 1996, DIME: A shared memory environment for distributed simulation, monitoring and control of urban traffic. In *Proceedings of European Simulation Symposium ESS'96*, Genoa, Vol. 1, pp. 152-156.

Bédard, Y. *et al.*, 1996, Adapting data models for the design of spatio-temporal databases. *Computer, Environment and Urban Systems*, Vol. 20(1) pp. 19-41.

Claramunt, C. and Thériault, M., 1995, Managing time in GIS: An event-oriented approach. In *Recent Advances in Temporal Databases*, J. Clifford and A. Tuzhilin (eds) (Berlin: Springer-Verlag), pp. 23-42.

Claramunt, C. and Thériault, M., 1996, Toward semantics for modelling spatio-temporal processes within GIS. In *Advances in GIS Research I*, M. J. Kraak and M. Molenaar (eds) (London: Taylor and Francis), pp. 27-43.

Claramunt, C., Parent, C. and Thériault, M., 1997, Design patterns for spatio-temporal processes. In *Searching for Semantics: Data Mining, Reverse Engineering*, S. Spaccapietra and F. Maryanski (eds) (London: Chapman & Hall), *in press*.

Clifford, J., Croker, A., Grandi, F. and Tuzhilin, A., 1995, On temporal grouping. In *Recent Advances in Temporal Databases*, J. Clifford and A. Tuzhilin (eds) (Berlin: Springer-Verlag), pp. 193-213.

Kosonen, I. and Pursula, M., 1995, Object-oriented and rule-based modelling - experiences from traffic signal simulation. In *Proceedings of the 6th International Conference on Computing in Civil and Building Engineering*, Berlin, Germany.

Oliveira, J. L., Pires, F. and Meideiros, C. B., 1997, An environment for modelling and design of geographic applications. *Geoinformatica*, Vol. 1(1), pp. 1-24.

Papazoglou, M. P., Laufmann, S. C. and Sellis, T. K., 1992, An organisational framework for co-operating intelligent information systems. *International Journal of* ICIS, Vol. 1(1), pp. 169-202.

Parent, C., Spaccapietra, S. and Zimanyi, E., 1997, Conceptual modeling for federated GIS over the Web. In *Proceedings of the ISPJ International Symposium on Information Systems and Technologies for Network Society,* Fukuoka, Japan, pp. 173-182.

Peuquet, D. J., 1994, It's about time: A conceptual framework for the representation of temporal dynamics in geographic information systems. *Annals of the Association of the American Geographers*, Vol. 84(3), 441-461.

Peytchev, E., Bargiela, A. and Gessing, R., 1996, A predictive macroscopic city traffic flows simulation model. In *Proceedings of European Simulation Symposium ESS'96*, Genoa, Vol. 2, pp. 38-42.

Snodgrass, R. T. *et al.*, 1995., *The TSQL2 Query Language*. The TSQL2 Language Design Committee (Kluwer Academic Publishers), 674 p.

Story, P.A. and Worboys, M., 1995, A design support environment for spatio-temporal database applications. In *Proceedings of the COSIT'95 Conference*, A. U. Frank and W. Kuhn (eds) (Berlin: Springer-Verlag), pp. 413-429.

Tansel, A. U. *et al.*, 1993, *Temporal Databases: Theory, Design and Implementation* (Benjamin Cummings).

Tryfona, N., Pfoser, D. and Hadzilacos, T., 1997, Modeling behaviour of geographic objects: an experience with the object modeling technique. In *Proceedings of the CASE'97 Conference*, Barcelona, pp. 347-359.

18

To interpolate and thence to model, or vice versa?

Claire H. Jarvis, Neil Stuart, Richard H. A. Baker and Derek Morgan

18.1 INTRODUCTION

Most environmental models take as their input a sequence of parameters that are observed at only a limited number of points. There are many practical circumstances in which one wishes to extend these models to make spatially continuous predictions. In such cases, the question arises whether to interpolate the inputs to, or outputs from a model into an output grid (Figure 18.1). The issue is important because the two approaches differ in the efficiency and quality of the results that are produced. With the number of situations in which dynamic environmental models are being linked to GIS now expanding rapidly, this is a generic and as yet under researched question that the GIS and environmental modelling communities need to tackle.

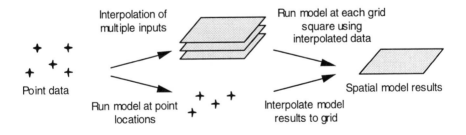

Figure 18.1 Interpolation of inputs to, or outputs from a model (After Burrough, 1992)

The question whether to interpolate or to model first arises on any occasion when a user needs to provide point data input to a model that is evaluated at multiple locations. In cell based GIS, the solution is usually first to interpolate the scattered points to fill the grid layers, then to run model computations on the multiple input grids (e.g. Aspinall and Pearson (1996)). In more recent work, a further temporal element has been introduced to this question by the linking of time-dependent point process or rule based models within a spatial setting (Burrough *et al.*, 1993). When incorporating time series of input point sets, the number of intermediate grids to be created by the *interpolate first* method is multiplied further by the number of days.

Many ecosystem models are driven by climatic variables that are highly dynamic over time (Cramer and Fischer, 1996) but for which data is sparsely distributed over space. If we wish to run an ecosystem model over space, do we really need to estimate how values of each input variable change in relation to the changing environment away from meteorological stations? Alternatively, might the ecosystem be equally well represented by running the model at locations only where the condition has been directly observed and then interpolating the results? This shows how the question arises practically in many applications where dynamic modelling requires spatial data inputs, including fire management, hydrological, crop yield, pest risk and nutrient transfer models (e.g. Kessell (1996), Landau (1996)).

The question posed has rarely been considered explicitly within the published literature. Burrough (1992), viewing the issue in terms of error propagation, suggested that the question could be solved using fully spatial multiple simulations. For applied scientists already frustrated by practical difficulties in the management of data and computational performance of current proprietary GIS when modelling in space-time (e.g. Johnston *et al.* (1996)), this may be considered impractical. Work by Heuvelink (1998) and Arbia *et al.* (1998) among others have progressed our ability to model error propagation within environmental models in space, but the temporal element in such processes has still to be addressed. For both practical and intellectual reasons therefore, applied studies often interpolate the outputs from, rather than the inputs to an environmental model. The application area for this case study, that of pest risk analysis, exemplifies this situation (e.g. Régnière *et al.* (1996)).

This work explores the interpolation-modelling issue for three models of varying complexity in the domain of insect ecology. Inevitably the structure of mathematical simulations within any case study are highly context dependent. Empirical studies are needed to allow us to quantify the magnitude of error that may be introduced into a single model result by choosing the less computationally intensive method of interpolating the point based outputs of a model. With many models working on sequences of daily or hourly data inputs, it is also vital to know if errors introduced at each time step could affect the logical coherence of these results if they are mapped out over time at a given place. The application area of insect ecology provides a realistic and thorough situation for testing, because rates of pest development may vary considerably over the landscape in both space and time. Models of insect ecology are complex in their treatment of processes acting over time in compared to many time-independent and rule based environmental models that have often been coupled with GIS.

Our analysis concentrates on quantitative errors within the attributes, rather than in position, as these are usually considered the more important component of error in environmental data analysis (e.g. Heuvelink (1998)). Focused numerical analysis is coupled with practical observations in order to provoke modellers into considering the broader implications of the two interpolation-modelling approaches. The aim is to avoid the intensive and time-consuming approach of multiple experimental simulations where possible and to assist researchers in evaluating the relative merits of performance versus quality issues in their own research domain. The integrity of results over time can be explored in this study because the models can be made to predict the date at which each pixel could reach a certain stage in an insect's phenological development. Three metrics are reported that might assist the applied user in determining whether to interpolate model inputs or outputs in a particular study. These include considerations of overall point based root mean square (r.m.s.) accuracy, logical errors in the developmental sequence of the insects and the spatial coherence of the interpolated results.

18.2 METHODOLOGY

While many factors influence pest populations (overall numbers), pest phenology (sequence of development) is in contrast relatively well understood. As a first step towards integrated insect/environment modelling incorporating dispersal processes, our goal was therefore to develop a system that can provide predictions of insect *phenology* across space. Such spatial phenologies are in themselves of practical value in two areas. Firstly, they are a useful source of crude measurements of establishment probabilities for non-indigenous pests (Baker, 1994) and secondly pesticide usage may be reduced by timing applications more carefully in relation to insect development. Previous examples of this type of work targeting agricultural or horticultural environments are few, although geospatial methodologies are increasingly being used in pest management systems within forestry settings (e.g. Leibhold *et al.* (1994)). Some American studies have focused on the interpolation of phenologies (e.g. Schaub *et al.* (1995)), but work using spatio-temporal inputs such as outlined in this chapter is still unusual.

Insect phenology models (e.g. Baker and Cohen (1985); Morgan (1992)) require, at minimum, the provision of sequential daily maximum and minimum temperatures. These variables form the model inputs referred to in this study. Three point process models, originally written to give estimates over time only, were used to investigate the consistency of the results. Often, little is known of the biology of an unusual invading species. Accumulated temperatures in Britain relative to those of the suspected country of origin can provide a first means of comparison. For this reason, the UK Meteorological Office model for accumulating temperature (Anon, 1969) was the first model chosen. Secondly, a generic phenological model (Baker and Cohen, 1985) was linked to provide more flexible modelling of non-indigenous species. In this chapter European parameters for the Colorado beetle (*Leptinotarsa decemlineata*), which poses a significant threat to British agriculture (Bartlett, 1980), were utilised. Thirdly, a relatively sophisticated non-linear boxcar model for codling moth (*Cydia pomonella*) (Morgan, 1992) was included to reflect the type of model more commonly used for insects indigenous to Britain.

As Figure 18.2 indicates, these three ecological models vary considerably in their conversion of data inputs (daily maximum and minimum temperatures) into pest development rates. At their simplest, development occurs in proportion to temperature accumulation whatever the stage within the pest life cycle the organism has reached. That is, the development rate is time independent. The highest degree of sophistication is represented by the sigmoidal development curves from the model for the development of the codling moth. In this case, development occurs at *pre-assigned non-linear rates* within each stage until reaching a standardised threshold upon which the pest cohort moves to a different stage in its life cycle, or expires. Graphically similar, the development of the Colorado beetle moves forward according to *pre-set development thresholds* that determine the development rates for each stage.

These process models were originally written to give estimates over time but with only point-based inputs and outputs. They have been adapted to provide predictions over

1km^2 neighbourhoods using partial thin plate spline (Hutchinson, 1991) and simple conventional voronoi interpolation methods. The system used to undertake this modelling is a prototype being developed by the authors for Central Science Laboratory (MAFF) for the purpose of undertaking spatial pest risk assessments. At present, the system is a loosely coupled suite of modules incorporating phenological models, interpolation routines, software for pre-processing UK Meteorological Office data files and the export of results in common GIS map formats. A graphical user interface, providing results in ArcView map/chart windows has recently been added to enhance its ease of use for more routine enquires (Gillick, 1998).

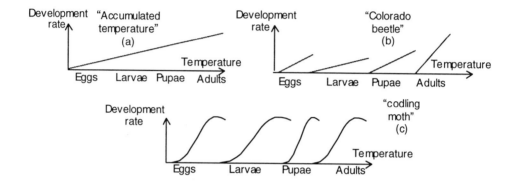

Figure 18.2   Relationships between development rate and temperature within the example phenological models

So that we might compare the effect of the order of the modelling and the interpolation steps upon the results, splining has been applied either to the input data prior to modelling, or to the results from modelling at data points. The former procedure is considerably more computationally intensive, since both the number of interpolations and number of model runs are increased. For each of the three models, a sequence of daily maximum and minimum temperature inputs throughout the growing season is required as input. Each model run is initiated at the start of a calendar year with a population of viable, mature adults. The process by which interpolator and covariates were selected for the interpolation of daily temperatures is reported elsewhere (Jarvis, 1998).

Given the scarcity of point temperature data relative to the spatial area interpolated, error estimation for the spatial models is carried out using a cross validation methodology (Efron, 1982). This iteratively drops one point out of the interpolation and uses this to estimate the error in the interpolation process. Computation of this measure within the software developed is efficient relative to many mainstream GIS since the process of estimating interpolation functions is separated from the grid production itself. This cross validation facility was used to generate a variety of error metrics from the results from each pest model by means of interpolating inputs (temperatures) and secondly by interpolating the outputs from the accumulated temperature model (degree days) and the phenological model (Julian dates). These are discussed individually below.

18.2.1 Overall root mean square error

In addition to tracking and reporting accumulated error at test locations at the end of the run period, the simulations were also stopped at significant stages within model runs to record the accumulation of error over time. In the case of the accumulated temperature model, runs were made over different base temperatures over time and results are presented for the r.m.s. error between observed and modelled accumulated temperature in °C over the run period. For insect phenologies, the model errors were reported as the difference in Julian dates at which particular points in the insect lifecycle are reached. Cross-validation by dropping out one data point in turn was used to compute r.m.s. error for interpolating model inputs (temperatures) and the error in interpolating model outputs (degree days or Julian dates).

18.2.2 Logical error in insect development sequence

Consideration of logical errors is more commonly made when checking categorical data or database integrity than when modelling processes (e.g. Lanter and Veregin (1992)). Given the necessity to follow the progression of insect development day-by-day at critical times of the year, preserving the correct biological sequence during a model run is just as important as accurately predicting when the insect is expected to reach a certain stage. We can be sure that if we run the interpolations of temperature and then run the biological model at each and every pixel, the model will enforce the correct sequence; temporal inconsistencies are not possible if modelling *after* interpolating. Logical errors will only arise when interpolating the Julian dates predicted at modelled points, over space. Errors in the date predicted for reaching a certain stage were computed by cross validation of results from interpolating phenologies at known points. The sequence of these dates was then checked for its logical consistency. Additionally, logical errors within the sequence of interpolated grids for the various stages of insect development were mapped for a small study area to investigate the spatial pattern of these errors.

18.2.3 Spatial coherence of the interpolated results

The results were also analysed to investigate how the spatial structure of the output grids produced by either interpolating or modelling first resembled the spatial structure of the known observations at sample points. Experimental variograms were constructed both of the estimates of accumulated temperature and of Julian dates at which each stage of development was reached for both the *modelling-first* and *interpolating-first* strategies. These were compared to a variogram computed using results calculated from actual values of temperature at the meteorological stations. The unit of lag used for computation was 20km, with 18km being the average nearest neighbour distance between the original meteorological data sites. The variograms produced were used to assess the quality of the output grids produced by the two techniques in terms of how their spatial smoothness or fragmentation matched that of the actual data. Where the chosen interpolation-modelling technique over-smoothes the model results, the variogram range of the modelled data might be expected to exceed that of the actual data. The reverse, less likely case occurs when the estimated range is relatively low. If no range is detected in the interpolated results but a range is clearly distinguishable in the actual variogram, then this implies that simple averaging techniques might perform as well as more sophisticated interpolation algorithms.

18.3 RESULTS

18.3.1 Overall root mean square error

Intuition tells us that errors associated with uncertain daily input values will accumulate through a model run. Errors in accumulated temperature may be expected to be smaller from runs over high base temperatures (Figure 18.3) simply because the number of days for which the base temperature is exceeded and therefore over which input errors are propagated will be fewer. The rapid increase in error when using voronoi interpolation may indicate that the propagation of errors is non-linear. From this we might infer that even small reductions in error gained when interpolating daily inputs are worth striving for. In this case, considerable care has been taken with the selection of covariates to guide the splining process, with an annual average error in the estimation of daily maximum temperature of 0.79°C and of 1.14°C for daily minima.

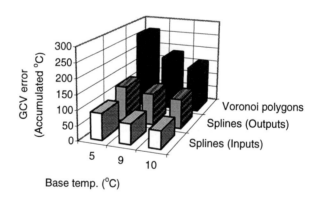

Figure 18.3   Cross validated error in accumulated temperature model results
(by interpolation technique and base temperature)

In relative terms however, for the accumulated temperature model (Figure 18.3), the codling moth model (Figure 18.4) and Colorado beetle (not pictured) the difference in terms of the GCV point based measure of accuracy at the withheld points between the two interpolation-modelling approaches is slight. For accumulated temperature this difference is approximately 30 degree days per annum for each base temperature modelled, while in terms of the date at which the phenological stages are estimated to be reached in the case of codling moth the average discrepancy from using the less intensive approach is only some three days. Current applied entomological practice for interpolating temperatures in simple entomological models is to apply voronoi polygons around meteorological sites. The poor accuracy of results computed using this method relative to those derived using partial thin plate splines is shown in Figures 18.3 and 18.4 for reference.

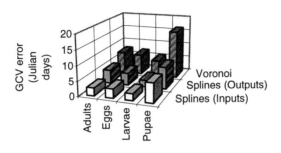

Figure 18.4 Cross validated error in Coding Moth phenology model results

Examining further the results for the codling moth model (Figure 18.4) the general effect of the time-dependent model structure causes the spatial error as represented by the cross validation statistics to rise over time. The dip in error during the larval stage may be explained with reference to Figure 18.2. Since the rate of development of the larval stage is slower than that of adults or eggs, the chances of predicting this stage correctly is correspondingly higher.

Given that differences between the predictions of date and degree days for the two interpolation techniques using partial thin plate splines are slim when averaged over England and Wales, the decision to interpolate model inputs or outputs will depend on the operational significance of the error bands and the variance of the individual point error values. The decision to increase the accuracy of predictions by interpolating daily temperatures might be justifiable where the efficacy of a chemical or biological application deteriorates rapidly.

18.3.2 Logical errors in time sequencing

Nationally, the proportion of landscape adversely affected by this sequential error is found to be surprisingly high. On the basis of cross validation statistics at test points for the Colorado beetle model, the dates at which each location reached a given stage suggest that the larval stage is erroneously predicted to predate that of the egg stage in 13% of cases, with pupae erroneously predating larvae in 22% of cases. Overall, one might infer that up to 24% of the England and Wales study area is affected by such problems, since the errors for each stage are not necessarily coincident. For the codling moth model, up to half of all test locations show a tendency to support the unlikely biological hypothesis that eggs are laid before the adults emerge from over-wintering.

Turning to the differences between gridded outputs for critical Julian dates, the sequences were checked and the spatial locations of the errors were mapped for the codling moth and Colorado beetle for 100km² in the Vale of York area (Figure 18.5). There is a strong tendency for errors in the logical sequence over time to occur at relatively high altitudes, where by the nature of the underlying distribution of UK meteorological sites, the interpolations are relatively poorly estimated. These areas also have rapidly changing environmental gradients, for example of temperatures, over short distances.

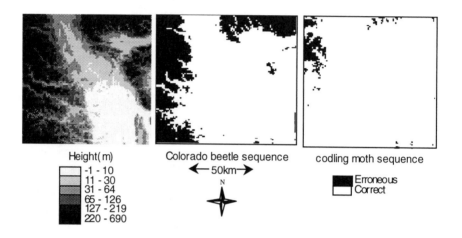

Figure 18.5 Errors in biological sequence created by interpolating phenological results, Vale of York, 1976

18.3.3 Spatial coherence of interpolated results

In addition to the way in which temporal sequences are represented by the two interpolation-modelling methodologies, the spatial correlation within the interpolated grids relative to the original data provides a further measure of the quality of the results. In the case of codling moth (not illustrated), variograms for the actual observed data and the grids produced by the two interpolated methodologies were of similar structure at all stages of development, reflecting the adequacy of both interpolation approaches.

Experimental semi-variograms for the Colorado beetle model reflected the greatest degree of spatial variability through time as the biological sequence progresses. At the early larval stage, little difference between techniques is seen (Figure 18.6(a)); either the interpolation of model inputs (temperature) or outputs (phenology) is warranted on this basis. However, beyond the pupal stage (e.g. immature adults, Figure 18.6(b)), the results created using interpolated phenologies suggest that spatial association occurs at distances of up to 140km. In contrast, those calculated using known inputs (temperatures at meteorological sites) suggest that little spatial auto-correlation is discernible. Surfaces for immature adults of the Colorado beetle created by interpolating the model outputs (phenological dates) will therefore appear grossly over-smoothed in comparison to those created using the interpolation of inputs (temperatures), which in this case should be preferred.

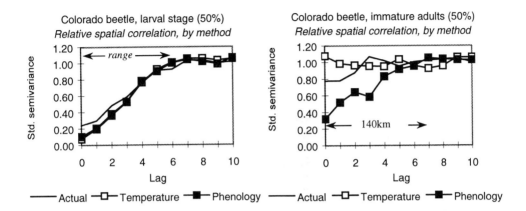

Figure 18.6 Experimental variograms for results of Colorado beetle (a) larvae and (b) immature adults computed using cross validated results from the two interpolation/modelling procedures in comparison with model runs made using actual data

For a 5°C base threshold temperature model, interpolating input temperatures also reflects better the actual accumulated temperature surface, as shown in the experimental variograms of Figure 18.7. The range of up to100km however suggests that overall degree of fragmentation is considerably lower relative to the results for the Colorado beetle model.

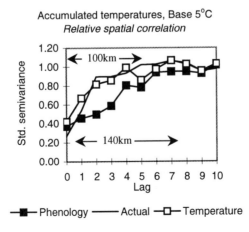

Figure 18.7 Experimental variograms for results of accumulated temperatures computed using cross validated results from the two interpolation/modelling procedures in comparison with model runs made using actual data

These results suggest that over short model runs, interpolating model outputs can produce adequate results. However, as time progresses the variance in the surfaces for Julian dates and for accumulated temperatures increases. This variogram analysis suggests that interpolating model outputs may produce gridded surfaces which are too smooth in comparison to those computed by interpolating inputs. As a result, at some point during the running of the model over a year, interpolating model outputs may become inappropriate.

18.4 DISCUSSION

Analysis of the surfaces created using the two approaches of interpolating temperature inputs or the modelled outputs (degree days or Julian dates), confirms the importance of considering carefully whether to interpolate model inputs or outputs. Despite the similar overall r.m.s. point accuracy of results obtained by modelling inputs and outputs, problems may arise with the logical sequence of output grids over time. For the modeller requiring a single output at the end of the time series this may not present a difficulty, while for those needing a sequence of outputs for further modelling such errors may be critical. The relevance of this error measure to the particular ecosystem under investigation may be gauged by mapping the phenological results, although it should be noted that this provides a relative, rather than absolute, measure of error. Levels of spatial fragmentation in the Colorado beetle results (Figure 18.6(b)) indicate that for long model runs using a model containing abrupt thresholds (Figure 18.2), the interpolation of input data may better reflect the spatial structure of the model results obtained at data points. Even where the overall point accuracy and semi-variogram analysis results between modelling inputs and outputs are similar, variance statistics show the interpolation of inputs rather than outputs to mimic more closely the underlying data.   Using all three metrics together, rather than the more traditional use of the aggregate r.m.s measure alone, provides a more rounded view of the spatio-temporal ramifications of the question posed.

The overall wisdom of considering the interpolation of model inputs rather than model outputs may also be considered in the light of this practical experience. In this case study, a number of factors pointed to the use of interpolated inputs, and at minimum a comparison of techniques. Firstly, the longer-term aim of the system under development is as a tool within a dynamic, integrated modelling environment. Predator, prey and crop models alike require the input of temperature amongst other variables. Furthermore the modelling of dispersion will involve multiple sequential surfaces, whichever interpolation technique is chosen. In the latter case, increases in computational intensity as a result of interpolating model inputs rather than outputs will relate to added model runs only, rather than both added model runs and interpolations.

Additionally, input data networks for other input variables required in the future, such as rainfall, do not necessarily coincide with those for temperature. To constrain modelling to points where all inputs are collected would reduce sampling densities to inappropriate levels. Finally, validating model output surfaces in the case of non-indigenous pests such as the Colorado beetle is impractical, whereas both input grids and the biological model can be independently tested.

The practical problems of spatio-temporal modelling within proprietary GIS favour the interpolation of model outputs (e.g. Johnston *et al.* (1996)). It may for example prove difficult to link the software with the high level code of the point process models, or to handle temporal sequences of interpolated data as anything but complete landscape grids. Obtaining cross validations will remain problematical while macro languages are computationally inefficient. The range of interpolation methods available within proprietary GIS software may not meet the needs of the sophisticated environmental modeller, whether it is one intending to interpolate either model inputs or outputs. In this case it may be that interpolation will be by public domain or in house code, as here. Then the barriers to interpolating model inputs rather than outputs relate more to computing power than to structural constraints. Our experiments indicate an approximate thirty-fold increase in computing time between interpolating model outputs versus interpolating input grids and then modelling. These issues are summarised within Table 18.1 as a question set that could be applied in other ecological contexts.

Table 18.1

	Interpolate inputs	Interpolate outputs
Limited computing power (PC environment)?		✓✓
Single output grid required?		✓✓
Smooth model results expected?		✓✓
Obvious gridded variables to guide interpolation of model results?		✓✓
Single input type?		✓
Input variables known to be difficult to interpolate accurately?		✓✓
Model results easy to validate?		✓✓
Input variables perceived to be spatially 'smooth' relative to model outputs?	✓	
Multiple input types with sample networks of different densities?	✓✓	
Desire for further integrated modelling using similar inputs in future?	✓✓	
Multiple sequential outputs required?	✓✓	
Model outputs difficult to verify?	✓✓	
Interpolated inputs of value in their own right?	✓✓	
Zero values in some output surfaces anticipated?	✓	
Multiple input variables process guided by different variables?	✓	
Complex linked model with abrupt thresholds?	✓	

✓✓ - Recommended, ✓ - Likely indication of best practice

18.5 CONCLUSIONS

Within an entomological setting, analysing the propagation of spatial and temporal errors in model predictions charts a new research area. Previous studies have interpolated results from phenological models unquestioningly, without incorporating checks on the spatial and temporal integrity of the process. The example of modelling the development of the Colorado beetle based on daily synoptic weather conditions shows that the common practice of interpolating the results of an at-a-point model will not *necessarily* produce results that are spatially and temporally coherent, even if point-based accuracy statistics (such as r.m.s.) seem satisfactory. In particular, logical inconsistencies in biological sequence may arise when interpolating phenologies that are avoided when interpolating inputs prior to running the phenological models. The interpolation of model inputs is also particularly beneficial for modelling the later stages of insect development, since variogram analysis of Julian dates based on actual data shows these to be highly fragmented.

Practical investigation of the question 'should we be interpolating inputs or outputs?' in relation to insect phenologies has resulted in a generic set of considerations that could be asked as part of the modelling process. These relate to issues of computing environment, model complexity and further use of the results. Considering such a question set does not require a large investment on the part of the modeller, who may be reluctant to delve initially into multiple simulations to solve the problem without necessarily seeing any direct benefits. There may however be no clear answer, and a number of focused simulations are recommended in such cases. This study demonstrates the value of three metrics designed for this purpose:

- Comparison of r.m.s. accuracies between interpolation methodologies at known points;
- Checks for logical consistency between temporal outputs;
- Analysis of relative semi-variance in results between models run using actual input data, interpolated input series and interpolated outputs.

More generally, the work highlights the need for further research to combine spatial and temporal error propagation methods and the need to understand the spatial significance of attribute error. Principles used to determine the integrity of spatial databases using logical rules may also be applied to spatio-temporal modelling procedures. The mathematical results for the particular phenological models exemplified are application specific. Nevertheless, the issues considered are generic to many other environmental models using time-varying inputs.

ACKNOWLEDGEMENTS

The provision of data by the UK Meteorological Office is gratefully acknowledged.

REFERENCES

Anonymous, 1969, Tables for the evaluation of daily values of accumulated temperature above and below 42°F from daily values of maximum and minimum temperature. *Meteorological Office leaflet*, 10, 10 pp.

Arbia, G., Griffith, D. and Haining, R., 1998, Error propagation modelling in raster GIS: overlay operations, *International Journal of Geographical Information Science*, Vol. 12, pp. 145-167.

Aspinall, R. J. and Pearson, D. M., 1996, Data quality and error analysis issues: GIS functions and environmental modelling, In *GIS and Environmental Modelling: Progress and Research Issues*, GIS World Books: Fort Collins, USA, pp35-38.

Baker, C. R. B. and Cohen, L. I., 1985, Further development of a computer model for simulating pest life cycles, *Bulletin OEPP/EPPO Bulletin*, 15, pp. 317-324.

Baker, R. H. A., 1994, The potential for geographical information systems in analysing the risks posed by exotic pests, In *Proceedings of the Brighton Crop Protection Conference - Pests and Diseases*, pp. 159-166.

Bartlett, P. W., 1980, Interception and eradication of Colorado Beetle in England and Wales, 1958-1977, *Bulletin OEPP/EPPO Bulletin,* 10, pp. 481-489.

Burrough, P. A., 1992 Development of intelligent geographical information systems, *International Journal of Geographical Information Systems*, Vol. 6, pp. 1-11.

Burrough, P. A., Van Rijn, R. and Rikken, M., 1993, Spatial data quality and error analysis issues: GIS functions and environmental modelling, In *GIS and Environmental Modelling: Progress and Research Issues*, GIS World Books: Fort Collins, USA, pp29-34,.

Cramer, W. and Fischer, A., 1996, Data requirements for global terrestrial ecosystem modelling, In Walker, B., Steffen, W., 1996, *Global Change and Terrestrial Ecosystems*, (Cambridge University Press: Cambridge), pp. 529-565.

Efron, B., 1982, *The Jacknife, the Bootstrap and other Resampling Plans*, S.I.A.M.: Philadelphia.

Gillick, M. W., 1998, GIS and pest risk assessment: issues in integration and user interface design for enhancing decision support. M.Sc. Dissertation, University of Edinburgh.

Grayson, R. B., Bloschl, G., Barling, R. D. and Moore, I. D., 1993, Progress, scale and constraints to hydrological modelling in GIS, In Kovar, K. and Nachtnebel, H. P. (Eds) *Application of Geographical Information Systems in Hydrology and Water Resources Management*, HydroGIS 1993, IAHS Publication No 211.

Heuvelink, G.B.M., 1998, Error *propagation in environmental modelling*, (Taylor and Francis: London), pp. 127.

Hutchinson. M. F., 1991, Climatic analyses in data sparse regions, In R. C. Muchow and J. A. Bellamy (eds), *Climatic Risk in Crop Production*, (CAB International, Wallingford), pp. 55-71.

Jarvis, C. H., 1998, The production of spatial weather data for the purpose of predicting crop pest phenologies, In Maracchi G., Gozzini, B. and Meneguzzo, F. (eds) *Proceedings of the Cost Seminar on Data Spatial Distribution in Meteorology*, 28 September - 3 October 1997, Volterra. *In press.*

Johnston, C., Cohen, Y. and Pastor, J., 1996, Modelling of spatially static and dynamic ecological processes, In *GIS and Environmental Modelling: Progress and Research Issues*, GIS World Books: Fort Collins, USA, pp. 149-154.

Kessell, S. R., 1996, The integration of empirical modelling, dynamic process modelling, visualisation, and GIS for bushfire decision support in Australia, In *GIS and Environmental Modelling: Progress and Research Issues*, GIS World Books: Fort Collins, USA, pp. 367-372.

Landau, S., 1996, A comparison of methods for climate data interpolation, in the context of yield predictions from winter wheat simulation models, *Aspects of Applied Biology,* Vol. 46, pp. 13-22.

Lanter, D. P. and Veregin, H., 1992, A research paradigm for propagating error in a layer-based GIS, *Photogrammetric Engineering and Remote Sensing*, Vol. 58, pp. 825-833.

Liebhold, A. M., Elmes, G. A., Halverson, J. A. and Quimby, J., 1994, Landscape characterization of forest susceptibility to gypsy moth defoliation, *Forest Science*, Vol. 40, pp. 18-29.

Morgan, D., 1992, Predicting the phenology of Lepidopteran pests in orchards of S.E. England, *Acta Phytopathologica et Entomologica Hungarica*, Vol. 27, pp. 473-477.

Régnière, J., Cooke, B., Lavigne, D. and Dickinson, R., 1996, A generalized approach to landscape-wide seasonality forecasting with temperature driven simulation models, *Environmental Entomology*, Vol. 5, pp. 869-881.

Schaub, L. P., Ravlin, F. W., Gray, D. R. and Logan, J. A., 1995, Landscape framework to predict phenological events for gypsy moth (Lepidoptera: Lymantriidae) management problems, *Environmental Entomology*, Vol. 24, pp. 10-18.

19

The application of GIS to the modelling of snow drift

Ross Purves, Jim Barton, William Mackaness and David Sugden

19.1 INTRODUCTION

Examples of the practical application of GIS to model processes in space through time, rather than to display static snapshots of processes in time, are remarkably rare (Worboys, 1995). However, spatio-temporal modelling of process is a common procedure in the environmental and physical sciences. This chapter describes the application of GIS to a spatio-temporal modelling problem, detailing the practical difficulties in using a 'pure' GIS approach. The inadequacies of GIS required the use of a loosely-coupled model utilising both a bespoke GIS and a high level programming language. The example focuses on the modelling of snow drift and discusses methodologies, results and an evaluation of the model. Finally we debate the need for GIS to drive environmental modelling approaches and vice-versa.

The model discussed in this chapter sought to describe snow drifting, and resulting zones of accumulation and erosion, over complex topography (mountains) during a storm cycle. Understanding snow drift is important for a variety of reasons (McClung and Schaerer, 1993):

- It is widely recognised as being an important factor in the development of windslab avalanches, where hard layers of snow deposited by the wind have good internal bonding, but very limited bonds with the adjacent layers in the snowpack;
- It impedes communications through closing roads and reducing visibility;
- It is an important factor in managing ski areas, particularly in Scotland where wind can redistribute large amounts of snow very rapidly;
- It has important hydrological implications as a result of differential distribution across catchments;
- It affects the survival of rare snow-loving plants dependent on sufficient accumulations during the winter.

Careful consideration should be given to the type of model used in seeking solutions to such problems. The model of snow drift developed was designed and tested on the complex topography shown in Figure 19.1, at Aonach Mor in the Western Highlands of Scotland.

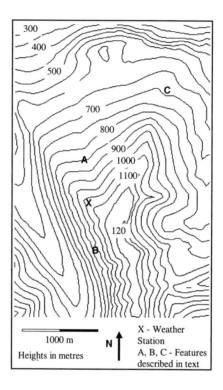

300
400
500
C
700
800
900
A
1000
1100'
X
120
B

1000 m
Heights in metres

N

X - Weather
Station
A, B, C - Features
described in text

Figure 19.1   The field site at Aonach Mor

19.1.1   Models of process

A number of different approaches could be taken to modelling snow drift, or indeed any other process.   For example, Burrough (1997) distinguishes between four types of modelling approach:

- rule-based (logical models);
- empirical (regression models);
- physical-deterministic (process-based, where in principle everything about a process is known); and
- physical-stochastic (the process is approximated by the model, but the probabilities are known).

It is often the case that models of real-world processes incorporate elements of all of these approaches, with the modeller attempting to use an approach that best describes and accounts for the specific processes involved. For example, there are very few examples of physical-deterministic models which completely describe a system, although we may be able to describe certain *elements* of a system physically. For instance, models of snow's response to short-wave radiation classically describe it in terms of Beer's Law (Oke, 1987):

$$Q_{Sz} = (1 - \alpha)Q_s e^{-\beta z} \qquad (19.1)$$

where Q_{sz} is the solar flux at a depth z;
         α is the albedo (reflectance) of the snowpack;
         β is the absorption co-efficient of the snowpack; and
         Q_s is the solar radiation incident at the surface.

This relationship describes an exponential dependence between incident radiation and radiation at a depth z, which is a well understood physical relationship. However snow's albedo and absorption are complex functions of wavelength, grain-size, density and impurities. An adequate physical model of these relationships does not exist, so these properties are generally described using some rule-engine based on empirical data collected in the field. A rule engine might give an empirically derived value to the albedo based on the length of time since the last snow fall or the air temperature over the preceding days, both parameters which are known to influence albedo. Hence, it can be seen that a single parameter is in fact described by three or four of Burrough's (1997) model classes.

Thus, in modelling a real-world environmental process the modeller must be prepared to embrace a variety of these model classes, in order to adequately describe the process concerned. The appropriateness of a particular model class requires us to consider both the spatial and temporal resolution of the model. For example snow drift may be described physically at an individual particle level over a short time period, but when describing snow drift over a mountain range or for the duration of a winter a more suitable aggregation of our understanding may involve using a rule-based or empirical model. Indeed a physically-based model may cease to be adequate as a result of the variability of the process in time and space and our inability to collect sufficiently detailed data either to parameterise or validate a physical model.

In this chapter we demonstrate how we have modelled snow drifting, drawing attention to the links with different modelling classes. Before we can construct our model, we must have some understanding of the processes which we wish to model, and in the following section a brief description of snow drifting processes is given.

19.2 SNOW DRIFTING AND WIND FIELDS OVER COMPLEX TOPOGRAPHY

Physically snow transport can be thought of as having three components: rolling, saltation and turbulent suspension (McClung and Schaerer, 1993). These processes have differing threshold wind velocities according to the properties of the snowpack in question (Schmidt, 1980). Rolling accounts for only a small proportion of snow in transport, with saltation and turbulent suspension being the main mechanisms of movement. Saltating particles bounce along the ground and are ejected from the snowpack and follow a trajectory, which may be up to a metre or so in height, before impacting back upon the snowpack. On impact these particles may in turn eject other particles from the snowpack. Since the particles being saltated store energy, it is possible for the wind speed to drop below the initial threshold value and particles may continue to be saltated. Finally, as wind speeds increase particles may be lifted into turbulent suspension. These particles may be lifted many 100s of metres above the ground, and many of them may be sublimated away before returning to the ground surface. In fact, in high winds, a large amount of the snow pack's mass may be removed in this way, resulting in an overall loss from the system. The flux of snow being transported is related to the wind speed and the propensity of a particular snowpack to drift. The propensity to drift has, in turn, a complex interdependence with the state of the snowpack, with for example, crystal size; snowpack wetness; bonding between crystals; and previous drifting, all having a bearing on the likelihood of snow drifting.

The wind speed and direction over the terrain in question is commonly referred to as the wind field and as a first approximation can be considered to be the gradient wind, that is the wind caused by the difference in pressure between two points, modified by Coriolis force. Thus over a plane surface the wind field could be considered constant at any instant in time over space. However over a complex topography this is not the case – the wind field varies over both space and time. A full physical model of the wind field over complex topography involves meeting mass and linear momentum balance conditions and having a full understanding of the initial conditions at the boundaries of the process. Most practical models of the wind field over complex topography make use of a number of empirical observations of the wind field such as wind speed reduction on leeward slopes and wind deflection when blowing across a slope (Oke, 1987).

19.3 THE SNOW DRIFT MODEL

19.3.1 Modelling the wind field

Snow drift is governed by the wind field. It is therefore important to derive a model that closely approximates a wind field over complex topography. There are a number of possible approaches to this:

- Application of a physically-based atmospheric model which satisfies all momentum and continuity equations (Piekle et al. 1992);
- Collection of extensive field data and interpolation of values between measured data;
- Collection of limited field data and utilisation of empirical relationships which relate the wind field to the topography.

In the context of this research the latter approach was the most practical in terms of deriving a computationally tractable model of the wind field over a large area and at high resolution.

The most obvious feature of a wind field in complex topography is the difference in wind speed at differing points across a ridge crest. We chose to implement a simple model of this, similar to the work of Bell and Quine (1996) who describe the use of a *shelter index* in defining wind throw hazard in forests. In our model we produced a map of wind speeds according to the aspect and inclination of a cell relative to the wind direction. GIS is ideally suited to such tasks where the process is being applied to a grid of cells *globally*. Each cell was examined, and assigned a value according to its aspect and inclination relative to the prevailing wind. If a cell lay in a swathe of aspects within 45 degrees of the mean aspect of a leeward slope, and its inclination was greater than 5 degrees then it was given an index, $0 < \text{Cell Index} <= 1$, according to the remapping function; the highest values lay directly in the lee, with the index decreasing towards zero when the slope's aspect was 45 degrees from the mean lee slope aspect. This value was then used to modify the wind speed for each cell. Thus on windward slopes no modification takes place, whilst on leeward slopes the wind speed is maximally reduced. The resultant grid was output, and stored as a lookup table for use within the snow drift model. Such grids were produced as required for model runs with different wind directions. This approach serves to illustrate a first order approximation of the reduction of wind speed in complex topography utilising a rule-based approach to describe how lee slopes are sheltered from the prevailing wind.

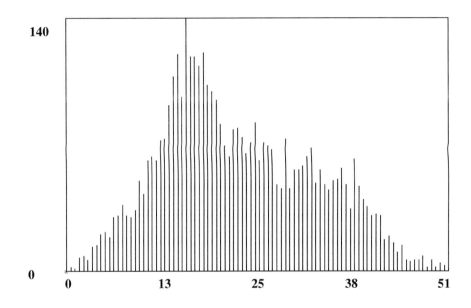

Figure 19.2 Histogram of slopes for 50 m DEM of terrain shown in Figure 19.1

It is known that topography produces a wind turning effect, and the next stage was to attempt to model such processes. Ryan (1977), developed a simple empirical equation for wind deflection in strong gradient winds blowing across a slope, where the inclination was less than 45°. Figure 19.2 shows a histogram of slopes for the 50m DEM used in the model, illustrating that this assumption is valid in using Ryan's equation for wind deflection:

$$F_d = -.225 * s_d * \sin(2(A-\theta)) \tag{19.2}$$

where F_d is the wind diversion [°],
          s_d is the slope gradient to the horizon in the downwind direction[%],
          A is the slope aspect [°] and
          θ is the wind direction [°].

This equation allows the wind direction in individual cells to be modified according to the aspect and inclination of a cell relative to the wind direction. The deflected vectors are calculated prior to creation of the shelter index. This is because Equation 19.2 modifies the wind direction and wind direction is subsequently used in the calculation of the shelter index.

By combining these quantities we generated a grid of wind vectors which had been locally modified by the topography. This model did not meet the laws of mass conservation, but did provide a simple methodology which could be utilised in describing the wind field in complex topography. The resulting patterns of the wind field reflect the underlying topography, with reduced wind speeds on lee slopes and wind vectors being turned by local variations in the topography (Figure 19.3).

Figure 19.3 Modified wind speed and direction for a 20 unit westerly wind. Lightest areas are minimum winds, arrows indicate wind direction

19.3.2 Modelling snow transport and accumulation

In our simplified description we chose to describe the flux of snow as a function of the wind velocity in the following way:

$$Q = k \, (u^3 - u^3_t) \qquad (where \; u_t > u) \tag{19.3}$$

where   Q is the amount of snow eroded from a cell,
         k is a constant,
         u is the wind velocity and
         u_t is the threshold wind velocity.

This relationship is empirically derived, and is based on work by Pomeroy (1993) in modelling the transport of snow over extensive, level snow covered surfaces.

The details of the procedure of moving snow around the grid were as follows. The wind vector at the grid's origin was examined, and the order in which the rows and columns were traversed was determined according to which component (x or y) of the wind vector had the greatest magnitude. This is an arbitrary choice, but some order for traversal of the grid must be chosen, preferably according to some logical rule. The movement of material was then modelled by reallocation from cell to cell, relative to the x and y components of the wind vector acting upon it, with the snow depth index (h) in cell $(x, t+\Delta t)$ after an iteration being found explicitly by the following equation:

$$h(x,t+\Delta t) = h(x,t) + Q(x,t) - Q(x+\Delta x,t) \qquad (19.4)$$

where Δt and Δx are the temporal and spatial steps respectively. This formulation means that snow erosion and accumulation are assumed to be instantaneous. Equation 19.4 is a finite difference equation, expressing a standard physical approximation for the change in depth over time.

The model was initialised assuming an equal distribution of snow over the whole topography. At the boundaries the flux was set to be constant, that is to say that the rate of erosion at the boundary cells equalled the rate of accumulation. This condition meant that material was available to drift in the model as long as conditions for drift were met at the boundaries.

19.3.3 Implementation of the model

The model was implemented using a combination of a bespoke GIS package (The GRID component of ESRI's Arc/Info) and a high-level programming language (Microsoft Visual Basic). This combination allowed us to take advantage of the spatial functionality of the GIS in undertaking static global tasks on the dataset. For example, the shelter indexes used in modifying wind speeds were generated by utilising GRID. Originally it was planned to develop the whole model within GRID. However, early experimentation demonstrated that GRID had very limited temporal functionality. The data structures used in GRID are designed for calculating changes in variables over an entire geographical space. Whilst this was ideal in calculating variables which were static through time but varied over space, for example the shelter index, such a paradigm is very poor at modelling changes in space and time, for example the movement of snow from cell to cell as a result of an imposed wind field.

A wind field was generated using a combination of GRID (to modify wind speed) and Visual Basic (to modify the wind direction). The resulting wind field was stored as an array, with a wind speed and a direction value for each cell. Material was then moved from cell to cell according to Equation 19.4 until a quasi-steady state was reached. At each time step the resulting distribution of snow over the terrain was displayed. It was possible to add more snow to the model over the entire area if required, and to vary the wind direction. The model was run on a 50m resolution Digital Elevation Model (DEM) over an area of some 15 km², as illustrated in Figure 19.1.

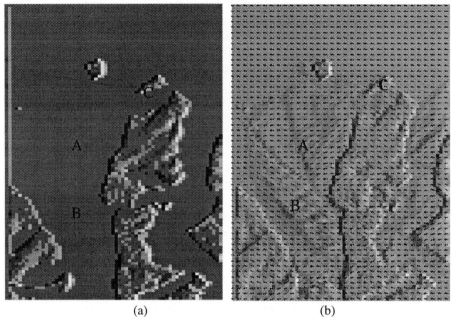

(a)   (b)

Figure 19.4   Relative snow depth (greatest in darker areas) with westerly wind after four iterations for:
(a) shelter index only, applied for westerly wind,
(b) both wind deflection and shelter index applied for westerly wind
A, **B**, and **C** are features described in text

19.3.4   Model results

Two sets of results are illustrated here to show the effect of the different components of the wind field modification. Figure 19.4a illustrates snow depth after four iterations under a westerly wind, where only speed has been modified. The eastern edge of the plateau, marked by the darker cells on the right of the figure, has the greatest accumulations of snow with the east facing corries generally showing greater accumulations. The windward cells in the accumulation zone lack snow, as a result of erosion, and the snow is trapped in the cells immediately to the leeward. In all other areas no modification of the snow depth occurs, and there is a constant flux into and out of the cells. However, if the model is left to iterate, the erosion zones slowly spread across the windward slopes until a quasi-steady state is reached.

In the second example, Figure 19.4b, the generated wind field is a combination of shelter index and deflected wind directions, again from an initially westerly direction. The same general pattern of accumulation can be seen on sheltered slopes, but a much more subtle pattern of shading is picked out on other areas of the topography. This pattern mimics the relief of the topography, with features such as Goose Gully (Feature A), which has a width of approximately the DEM's resolution, clearly identifiable as areas of increased accumulation. The striated effect on the western side of the mountain (Feature B) shows the effects of the various deeply incised gullies on this face.

The results clearly show areas of accumulation in sheltered zones, as one might expect. The most notable feature in the first example (Figure 19.4a) is feature C. This is the new beginner's ski area at Nevis Range, and was chosen as a good snow holding site under the prevailing conditions (west and south-westerly winds). It is a relatively gentle depression, which holds snow better than the surrounding areas due to its increased shelter from the prevailing winds. Also well defined is the break in slope from the plateau to the eastern corries of Aonach Mor, which is clearly shown to collect large quantities of snow.

In the second example, Figure 19.4b, Goose Gully (Feature A), the most prominent natural snow holding feature on the north face of the mountain, holds snow preferentially compared to the surrounding slopes, as a result of the effects of wind deflection in the model. Likewise the gullies on the western side of Aonach Mor are shown to collect snow, even under westerly winds. The wind deflection heuristic allows gullies to collect snow preferentially by angling the wind vectors in towards the gully, and thus causing a net gain of material.

19.4 DATA COLLECTION FOR VALIDATION AND CALIBRATION

A range of meteorological, snowpack and topographic data were necessary in order to run, calibrate and validate the model (Table 19.1). Our planned strategy was two-fold: to collect data to be utilised in the model's rule-engine, and data to validate the model's output. Examples of data to use in the model's rule-engine included the following:

- threshold velocities for snow drift under differing conditions;
- threshold velocities for differing snow transport processes;
- driftability measurements as a result of conditions imposed upon a snowpack; and
- flux measurements at different heights as a function of velocity.

Threshold velocities were measured using a hand-held anemometer and observations of snowpack data. The driftability data were collected using a combination of observations and SAIS (Scottish Avalanche Information Service) data. Finally the flux was measured using instruments modelled on the driftometers developed by Robert Bolognesi of the SLF (Swiss Federal Institute of Snow and Avalanches). These consisted of nylon socks mounted on plastic piping at varying heights. Flux was measured by making timed measurements of weight of snow collected in the socks. They provide an index of the rate of flux rather than an absolute measurement and functioned successfully in winds up to around 20 ms^{-1}.

Field work was undertaken at Nevisrange ski area, on Aonach Mor, near Fort William in the Western Highlands (Figure 19.1). Very limited snow cover existed and both winters were characterised by one or two snow falls in conditions which were very stormy, followed by little fresh snow for the rest of the winter. As a result, work on validating the model by measurements in the field was not possible. Validation of the model relied instead upon the local knowledge of the SAIS avalanche forecasters and examination of Nevisrange's considerable collection of photographs of the ski area under different conditions. Although this situation is less than ideal, we believe that given the conditions it represents the best possible compromise. The features described in the results section were found to be in good agreement with both what SAIS observers would expect under such conditions, and furthermore to closely follow areas chosen as pistes due to their natural snow holding ability.

Table19.1   Data collected during project

Data type	Resolution/ Measurement Unit	Source	Purpose
Digital Elevation Model (heights)	50m/ m	Ordnance Survey	Topography on which model was run
Wind speed and direction	Point measurements, continuous /ms^{-1}, degrees	Met. Office, ski area and portable Automatic Weather Stations (AWS)	Model initialisation and validation
Snowpack properties	Point measurements, daily/ varied	Scottish Avalanche Information Service daily measurements and fieldwork	Model initialisation and validation
Snow drift measurements	Point measurements, continuous/ varied	Purpose built instrumentation and field work	Model calibration

19.5   DISCUSSION OF IMPLEMENTATION OF MODELS IN GIS

This chapter has demonstrated a spatio-temporal model of a real process, namely snow drifting. It has utilised physical, empirical and rule-based approaches to develop a model of a complex process, which gives results which are qualitatively in good agreement with observed behaviours. The model was implemented using a loose linkage between a GIS and a high level programming language. The authors would argue that such an approach, contrary to that often argued for in GIS (Burrough, 1996; Fedra, 1993), makes the best use by combining applications to provide an appropriate solution.

The lack of practical examples of spatio-temporal processes implemented within bespoke GIS is perhaps illustrative of the fact that GIS is unsuited to the task. Although add-ons to GIS may claim to give further temporal functionality, we would argue that the modeller requires a high level of transparency in programming such iterative processes and as such the modeller is likely to prefer a high level programming language. However, as shown in this chapter GIS is extremely well suited to modelling spatial processes on both a local and a global scale. Furthermore, it can greatly aid the modeller in visualising the results of a model at different time steps. Current GIS do not lend themselves to loose-linkages however, since the extraction or input of data tends to require manual intervention. It is our contention, that rather than try to include temporal elements in GIS suitable for spatio-temporal modelling (a development which is GIS driven) developers should concentrate on mechanisms for data transfer or data serving which allow GIS to be a useful addition to the modellers toolkit, rather than attempting to replace that toolkit entirely.

19.6 CONCLUSIONS

This chapter has sought to debate two interrelated issues. The first centres around the question of appropriateness of models, reminding us that the choice of model should be driven by the research question at hand and not simply by a desire to use GIS. We demonstrate clearly how real-world models encompass a range of model classes according to both our understanding of the process being modelled, the spatial and temporal resolution of the process being modelled, our ability to validate the model and the level of detail required in the solution.

The second element of this chapter has demonstrated these ideas through a qualitative model of snow drifting over complex terrain. Through a suitable combination of GIS and a high-level programming language we developed a loosely linked spatio-temporal model which produces results which match up well with expectations. The model uses a number of simple concepts, such as shelter index, wind deflection and driftability which are suited to further development. Its modular nature means that the refinement of the rules to produce a quantitative model is possible, building on the framework that has been developed. Although conditions were poor, and full validation was not possible, we developed a suite of techniques suitable for monitoring snowdrift in the harsh Scottish climate.

ACKNOWLEDGEMENTS

The assistance of Nevis Range Development Company, Graham Moss, Mark Hughes and Blyth Wright of the SAIS is gratefully acknowledged. Andy Kerr is thanked for his valuable comments on the draft manuscript. This research was funded by NERC grant GR9 / 2053.

REFERENCES

Bell, P. D. and Quine, C. P., 1996, Calculating an index of wind damage for British forests using GIS. In *Proceedings of the GIS Research UK 1996 Conference*, Canterbury, UK, Association of Geographic Information, pp. 199-202.

Burrough, P. A., 1997. Environmental modelling with geographical information systems. In *Innovations in GIS 4*, Kemp, Z. (ed) (London: Taylor and Francis). pp 143-153.

Fedra, K., 1993, GIS and environmental modelling. In *Environmental modelling with GIS*, Goodchild, M. F., Parks, B. O. and Steyaert L. T. (eds) (Oxford: Oxford University Press), pp 35-49.

McClung, D. and Schaerer, P., 1993, *The avalanche handbook*. (Leicester: Cordee).

Oke, T. R. 1987., *Boundary layer climates*. (London: Methuen and Co.).

Piekle, R. and 10 others., 1992, A comprehensive meteorological modelling system RAMS. *Meteorological Atmospheric Physics,* Vol. 49, pp. 69-91.

Pomeroy, J. W., Gray, D. M. and Landine, P. D., 1993, The Prairie blowing snow model: characteristics, validation and operation. *Journal of Hydrology*, Vol. 144, pp. 165-192.

Ryan, B. C., 1977, A mathematical model for diagnosis and prediction of surface winds in mountainous terrain. *Journal of Applied Meteorology*, Vol. 16(6), pp. 571-584.

Schmidt, R. A., 1980, Threshold wind speeds and elastic impact in snow transport. *Journal of Glaciology*, Vol. 26(94), pp. 453-467.

Worboys, M. F., 1995, *GIS: A computing perspective*. (London: Taylor and Francis).

20

Modelling river floodplain inundation in space and time

Carsten Peter and Neil Stuart

20.1 INTRODUCTION

Within the broad range of environmental models presently developed, certain process models use parameters that increasingly in future will be derived from digital geographies held in GIS. We use the term *spatial process model* to distinguish those environmental models which have a strong element of spatial representation within the model structure. Some examples are models of soil erosion, surface insolation or water runoff which are strongly influenced by terrain variables, or models of animal or plant dispersal where the dispersal process is strongly influenced by the spatial pattern of land use within an area. Many examples have been reported of these kind of environmental process models being linked to GIS to permit practical enquiries, problem solving and hypothesis formulation in a diversity of environmental sciences (see for example, Moore *et al.* (1993), Goodchild *et al.* (1996), Kovar and Nachtnebel (1996), Johnston (1998)).

Although GIS appear to be an obvious platform on which to run spatial process models, several practical attempts to link existing process models to GIS have reported problems. These can be traced in part to mismatches between how spatial properties are conceived in the process model, compared with their representation within the spatial data models of GIS (Raper and Livingstone, 1994). Some specific difficulties are that GIS presently lack data structures for conveniently handling data for multiple time steps and the internal programming languages for GIS are usually less flexible than standard programming languages for creating the data structures needed for process modelling (Wesseling *et al.*, 1996; Burrough, 1997).

The majority of environmental modelling with GIS to date has been with pre-existing process models, linked through one of a range of methods described by Fedra (1996) as ranging from a shallow coupling through to a deeper integration. Whilst some of the difficulties of linking models and GIS result from the limitations of GIS, there are also problems because most of the models which have been linked to GIS in the last few years have been in existence for considerably longer. Consequently, one is often inheriting code written for an earlier generation of computing equipment and from a time when the amount of environmental information in digital form was orders of magnitude less than today. Whilst the idea of feeding an established, but old model with richer data may seem a consistent and a pragmatic solution, recent research is revealing that most models are tuned to a particular level of generalisation in their environmental data inputs. Models are often designed to implicitly model processes with some degree of spatial averaging – for example to estimate the runoff from a whole valley side. If greater detail is required, the model may need to be re-formulated to represent processes acting at that greater level of detail. This implies recalibration, possibly against more spatially disaggregated observations. Taken together these constraints may mean that the limits of an old model have been reached and an alternative approach to modelling the process needs to be developed (Grayson *et al.*, 1993).

In a comprehensive review of the state of the art in modelling environmental processes Wilson (1996) concludes that the present technology for the collection and manipulation of environmental data at higher resolution provides many new opportunities for the development of new models. In the rest of this chapter, we illustrate how enhanced models of floodwater inundation processes may be developed for GIS. Novel aspects of the approach are how a relatively simple formulation of water movement, when combined with a more detailed and structured representation of the terrain, can at certain scales provide plausible predictions of the depths and extents of ponding over an area.

Particularly in the context of providing operational systems for decision support, Fedra (1996) identified the high degree of flexibility that can be achieved from generic, low-level building blocks, in an object-oriented design. This concept will be exemplified in this study, where per-pixel digital terrain analyses are the building blocks used to compute the geometry and connectivity of surface depressions. Once computed, by then recognising these depressions as higher level objects, one is able to model the time-dependent movement of floodwater between depressions in a flexible and efficient manner. Since the depression objects retain a definition of their constituent elevation cells, as well as information on the current volume of water input, maps such as the depth of ponding in a depression can be produced on demand for any time step. The modular software methods developed here are designed to be adapted for use on new and more spatially detailed elevation data becoming available from large scale photogrammetry, LIDAR altimetry and SAR interferometry for floodplain areas.

In the following sections, the modelling concepts and assumptions are explained, with particular emphasis given to the extensions to existing GIS data structures that were required to support the modelling of the inundation and drainage processes over multiple cycles of flooding. This is followed by some example results, where the methods were used to simulate the extent and duration of flooding into low-lying ground by the tidal river Thames, resulting from a hypothetical breach of a river bank during surge tide conditions.

20.2 FLOODWATER INUNDATION - EXAMPLE OF A SPATIAL PROCESS

We present a simple example to illustrate the opportunities presently afforded for undertaking spatial process modelling using new digital data and which begins to extend the present data structures of GIS to support more effective manipulation and analysis of spatio-temporal data.

The example chosen is to model the inundation of a low-lying flood plain area in the event of a river bank failure during a surge tide. A hydrological example has both research and practical relevance.

At a research level, hydrological analysis is often cited as one of the most developed area of GIS functionality for environmental modelling. This study will show how far a relatively simplified hydrological problem can be analysed using the data structures and functions of current GIS.

At a practical level, the flooding of the Rhine, the Mississippi and the River Oder in recent years have shown that even in industrialised countries where millions have been spent on flood protection, floods remain a major threat to people and property.

The research undertaken to predict these floods has shown that although hydraulic and stochastic river models can predict stage levels with a certain degree of accuracy, this is not necessarily the information that is needed by the 'end-user' in order to manage and respond to the flood risk. The information needed when planning the response to a flood risk includes answers to questions such as:

- How deep will the water inundation be on the ground?
- Which areas will be flooded and when?
- What types and extents of damage will this produce?

The process of re-writing existing hydrologic models to exploit higher resolution spatial data is a significant task and the outcome may be that a new model formulation is required. An opportunity therefore exists to develop methods that more explicitly use spatial representations of the main influences on hydrological processes and are more geographic in portraying modelled results.

20.2.1 Using GIS to model floodwater inundation

The present speed of information processing creates opportunities for real-time forecasting of floods. The prospect of modelling flood events in faster than real time and of combining modelled results for flood depths with other information in a GIS provides a basis for planning and managing the response of emergency services (Roberts and Anderson, 1994; Gatrell and Vincent, 1990).

However, hydraulic models require large sets of input data (hydraulic characteristics such as roughness coefficients, channel dimensions, etc.) which still have to be acquired by field surveys. These models are computationally intensive and complex so that setting them up for a specific scenario is costly.

A breach in a riverbank, for example, can happen in a very large number of places along a river's course. It is neither practical nor economic to set up and run a hydraulic model for every possible breach location, yet planners need to have formulated a response for the possibility of a breach at a large number of candidate locations. There is also the need to respond in faster than real time to an actual event and to gain some understanding of the areas likely to flood in the next few hours.

This leads us to formulate a model which is relatively simplistic in its assumptions about the detailed physical processes of flooding, but which exploits higher resolution terrain data to predict where surface water will accumulate once a breach has occurred. The model needs to accept the location of the breach from the user and to produce predictions in the form of maps showing the depths and duration with which different areas are expected to be inundated. The following sections provide more detail of how a software module was developed using this approach to flood modelling and how results can more easily be integrated with other relevant information within an existing GIS.

20.2.2 Modelling concepts

The model describes the inundation of a floodplain following the failure of a river embankment at a location specified by the user, perhaps as the result of receiving a report from a field officer. The model is designed to use input data that is likely to be already available in a GIS without the need for any special field survey. A breach flow model calculates the inflow and outflow through a breach or floodgate given the estimated dimensions of the breach and the water level on both sides of the failure. The approach used for the test study is a standard broad-crested weir formula with submergence correction factors. A more sophisticated breach flow model can easily be used if necessary.

The floodplain is conceptualised as a series of depressions that act as storage cells for floodwater. In traditional quasi-2D flood modelling (Cunge et al., 1980), the sequence and connections between depressions must be pre-defined before the model is run. As such, this type of modelling makes no explicit use of spatial data during its execution. Our proposed model does not require such prior definitions of terrain. The sequence and connection of depressions is determined automatically by processing data derived from a digital elevation model using methods developed by, amongst others, Jensen and Dominigue (1988). Such methods are found increasingly within GIS (Moore et al., 1993).

20.2.3 Assumptions of the model

The model that has been initially implemented makes the following assumptions:

- The model does not calculate the time it will take for water to travel from the breach or overflow point to the lowest point of the next depression. Given that the distances are small (some 100 m) in all cases but for very large, shallow depressions the time it will take the water to travel to the lowest point can be assumed to be small in comparison to the time it takes to fill the depression (Evans and von Lany, 1983).
- The distribution of excess water from one depression to another is assumed to happen instantaneously, as soon as one depression is full to capacity. Hydraulic functions such as weir formulas are not used to model the flow between depressions. This ignores the possibility of water reaching a higher level in a depression than the level of the outflow point, while water is entering faster than it can be transmitted to the next depression – an effect known as lake storage. Neglecting lake storage means that the maximum water levels in pools are slightly underestimated, while the flooding of the following depression is likely to be slightly faster than if this effect was considered.
- In this first model, water can only enter or leave the terrain via overland flow. Flow through tunnels and the sewage system, infiltration and groundwater flow are not modelled. The effect of this assumption is that a reduction in ponding levels due to infiltration or drainage into sewers is neglected. Also flooding through underground flow (e.g. water entering a storage depression through a tunnel) can not be predicted.

For this initial evaluation, where the focus is mainly on developing the digital terrain analysis procedures, these simplifications seem justifiable. The combined effect of these assumptions and simplifications for the flooding process will be to slightly under-predict maximum flood level of a depression and for flooding to occur slightly sooner. Given the envisaged use of such models for planning evacuation of areas, predicting a worst-case scenario for a given set of conditions is effectively erring on the side of caution. The model structure has been designed to allow refinements such as the incorporation of hydraulic linkage functions between cells and a sub-model to describe infiltration.

20.3 DATA AND METHODS

20.3.1 Data pre-processing

The algorithm that spreads a volume of water from an input location over the terrain requires as input three raster files:

1. an elevation model of the area under consideration;
2. the local drainage direction for each pixel in the elevation model;
3. the arrangement of pixels into watersheds throughout the elevation model.

The DEM is the only primary data set required to begin modelling. Data sets (2) and (3) are in turn derived from (1) by functions which are increasingly found within raster GIS. Computing the pixel memberships of watersheds (depressions) before execution of the flooding algorithm is not absolutely necessary, but is desirable in practice as it substantially reduces the processing time.

The public domain GIS, PC-RASTER (Deursen and Wesseling, 1993) was used for the pre-processing of the raster DEM. PC-RASTER was chosen as it offers advanced options for the flexible recognition of depressions in a terrain model. Some of the algorithms were developed to study how water ponds in surface depressions to form seasonal lakes on the plains of Alberta (MacMillan *et al.*, 1993). This distinguishes this package from most other watershed analysis tools that merely treat depressions as artefacts that are automatically filled in order to remove them and to permit a continuous river network to be derived from a DTM (Jensen and Dominigue, 1988).

The result of the first pre-processing step is a local drain direction (LDD) matrix where every pixel is assigned a number indicating the direction in which water would drain to one of its eight neighbours. Cells that only have neighbours at a higher or the same elevation are termed *pit cells*. These cells are local minima, having only neighbours whose drain direction is pointing towards them. In a second step all internal watersheds are found in the area under consideration. An iterative algorithm starts at every pit cell and follows up the drain direction for every possible path until it hits a cell that drains away to another pit. All the cells found are marked with a watershed identification number that relates the watershed to its pit cell. When the algorithm is finished, every pixel in the area has been allocated one of these watershed identification numbers. Although these internal drainage areas are identified using modified software for delineating watersheds, these are much smaller areas, defined by much more localised micro-topography compared to the extensive area that is typically thought of as a watershed. The term *depression* is more appropriate for describing these local terrain minima, since in most cases all of these depressions would be contained entirely within the watershed of one small river.

20.3.2 The FLOODFILL algorithm for terrain inundation

With these data sets derived, the user needs only to choose a function for controlling the volume of water input and to use a mouse click to specify a location for the water input (breach). The FLOODFILL algorithm then takes water from the breach and spreads this over the terrain. It recognises when and where water flows over into another depression, when a rising water level leads to pools merging to form one larger pond, and when this pond will overflow again into another depression, etc. This sequence of terrain filling is stored and used for modelling the later drainage of the terrain, and to account consistently for re-flooding of depressions that are already partially or wholly full of water, during a subsequent cycle of flooding.

20.3.2.1 Flow of processing

First, the algorithm determines the number of the depression in which the water input or outflow (i.e. the breach) is located. All pixels belonging to this depression are then added to a pixel list and sorted by their height value with the pixel of lowest elevation being first in the list. This pixel is the pit-point of the depression, using PC-RASTER terminology. An object of type *depression* is dynamically constructed. The data structure of this object is shown in Table 20.1.

Table 20.1 Data structure of the object *depression*

Name of data element	Role of this data element
PixelList	Cell location and elevation of all pixels in this depression (sorted with lowest pixel first)
Overflow	stores the overflow height to the next depression
WaterLevel	stores the present level of water ponded in the depression
PixelArea	area of one pixel
Pointer	moves up the pixel list for water volume calculations
Prev Pointer	points to the previous depression
Next Pointer	points to the next depression
Next2Drain Pointer	points to the depression that is the next to be drained

For each time step, the algorithm gets a volume of water as input and starts spreading this volume over the first depression. As Figure 20.1 illustrates, computing whether the input volume can be stored within the depression involves moving a pointer up the pixel list.

The pixel area is assumed to be 1 sq. unit for the formulae in this figure. By having a map of all depressions derived from the watershed analysis of the DEM, pixels are pre-sorted by elevation and by depression number. Within a single depression, this ensures that every pixel with a certain elevation value can be assumed to be flooded if the water level in the depression reaches that elevation. Without this pre-processing step, the modelling of the spreading of water would involve intensive neighbourhood operations to find adjacent pixels under a certain elevation value.

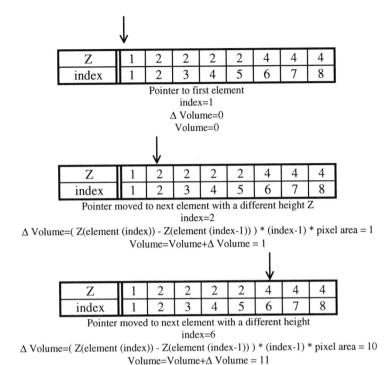

Figure 20.1 Volume calculation

The pointer moves up the sorted list and counts pixels until it reaches a pixel with a higher elevation. The floodable volume that fits between the terrain surface and the next higher elevation level is determined by the number of pixels with an elevation smaller than this higher elevation, multiplied by the area per cell times the elevation difference. These floodable volumes can be accumulated as a series of incremental slices, by advancing the pointer to successively higher elevation levels until the complete volume of input water in this time step has been distributed. The pixels are then sorted back into their original row and column order, so an output raster can be written for this time step, either by marking all the flooded pixels with a certain value or the values of the water level above some arbitrary ground datum. This part is relatively straightforward as long as there is only one depression. Usually, however, there are multiple depressions and the first will only fill up to a certain level, after which the water will then overflow the boundary and start flooding into another depression. It is therefore necessary to test for every step up the sorted list whether the pixels that are to be marked as flooded have a neighbouring pixel in another depression. If this is the case, the water level has reached a depression boundary. Because of the discretisation of elevation values in a raster DEM, it is not necessarily the elevation on this side of the boundary that determines the overflow height. All pixels on this and the other side of the boundary have to be checked and the higher value used as the minimum overflow height.

Figure 20.2 illustrates the construction of multiple depression objects and the work of the pointer chain in maintaining the correct relationship between depressions at each stage during flooding. Figure 20.3 illustrates the use of pointers to track the drainage of depressions. Both figures assume water is flowing in or out of depression 1 from a breach that must be imagined to be behind the plane of the paper.

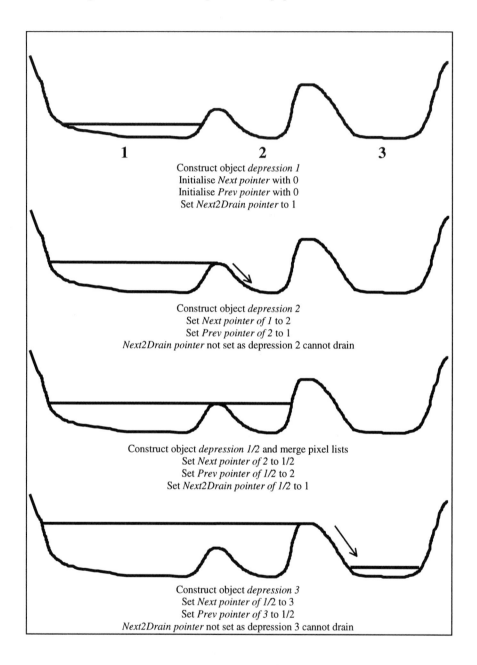

Figure 20.2 Flooding depressions in the terrain

The inflow process

Once the water level reaches the overflow elevation in the first depression, a new *Depression* object is constructed, with its *Previous* pointer set to point to the first depression. The elevation at which overflow occurred is stored in the data structure of the first depression and the *Next* pointer of the first depression is set to point to the newly constructed depression object (Figure 20.2). As soon as the overflow elevation is reached, the algorithm redirects any remaining volume of water from this time step into the new depression. From now on, the water level in the first depression remains constant. A new pixel list is generated and sorted and the steps above are repeated for the second depression.

At a further stage of processing, the second depression may overflow into the next by an overflow point with an elevation that is lower than the overflow point from the first depression, in which case the process is again repeated. It is also possible that the water in the second depression fills up to the height of the overflow point from the first depression. As soon as this water level is reached, the first and second depressions merge and have to be treated as one single, larger depression. This is done by creating a new depression object, whose pixel list is produced by merging and re-sorting the first and second pixel lists. The water level in the two merged depressions now increases as if they were one and may eventually overflow into another depression, and so on.

The *Next2Drain* pointer will only be set to the newly constructed depression if the first depression (the one with the breach) is part of the newly constructed one. Only then could this depression partly drain by return flow under gravity, as water levels in depression 1 are reduced (Figure 20.3). The *Next2Drain* pointer chain is important for representing the reverse process of outflow from the terrain, which is considered next.

The outflow process

After a certain interval, the flood level will fall below the height of the breach and water will begin to flow by gravity out of the depressions and back through the breach. Feeding the FLOODFILL algorithm with negative volumes of input water simulates these drainage processes. The algorithm starts with the current *Next2Drain depression* (presently *depression 1/2* in the example). The outflow volume is subtracted from the water that remains in this depression until the water level reduces to the height at which overflow occurred from the previous depression. The *Next2Drain* pointer chain is then followed back to the first depression and water is then removed from this. Only depressions that can eventually drain by gravity flow are in this pointer chain (Figure 20.3). Calculating how the water level will fall as water drains from the depression involves going through the pixel list in the reverse direction and calculating the incremental volumes removed for each elevation 'slice' using the method described above.

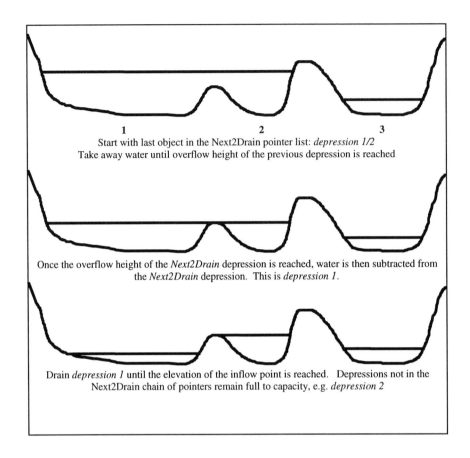

Figure 20.3 Outflow process

A larger pond, where depressions have been merged, will only drain down to the elevation of the highest overflow point amongst the depressions that comprise it. If the water level falls below this elevation, then the water in the highest depression becomes disconnected hydrologically from the ponds with lower overflow elevations. As the water level continues to fall, a series of surface ponds are left in the individual depressions, full to the brim, without an outflow. Algorithmically, this involves checking the water level in the depression that is presently being drained against the overflow elevation of the depression pointed to by the *Next2Drain* pointer. Once the water level reduces to this elevation, this is stored as the static water level in the current depression and the operations to remove water volumes are then applied to the *Next2Drain* depression. The process ends when the water level in the first depression reduces to the elevation of the breach, or the requested volume has drained from the terrain.

The *Next* and *Previous* pointers are used for the visualisation of the results and for subsequent re-flooding of previously inundated terrain. All the depressions that have received floodwater remain in memory. Some of the merged depressions and the first depression nearest to the breach will have drained down below capacity and the depression that was last to receive water may also be only partly full. The other individual depressions are assumed to remain at their full capacity after being disconnected by falling water levels. During the drainage process, as each pond becomes disconnected, its water level is set to the overflow elevation of the *Previous* depression. This is also the water level at which merged, composite depressions revert back to a series of two or more individual depressions, obtained by following *Previous* pointers. During a second cycle of floodwater input, the *Next* pointer chain is followed. As soon as the first depression is filled back up to its overflow, the *Next* depression and the volume required to re-fill it can be quickly calculated.

20.4 RESULTS FROM A SIMULATED BREACH ON THE TIDAL RIVER THAMES

The model has been applied to a scenario set by the Environment Agency to investigate in detail the inundation predicted to result from a hypothetical 75 m wide breach in a floodwall of an embayment of the River Thames. The location chosen was near to Barking and Dagenham, downstream of the Thames Barrier in central London. River levels for this hypothetical event were calculated using the curve for a recent tide that resulted in a closure of the Thames Barrier. The surge component of this tide was then increased to reach the peak tidal level of the 1953 London flood. Inundation depths over space and time were produced and combined with digital map data to give a geographical impression of the predicted flood impact. A series of inundation maps were produced at 20 minute time intervals to investigate the effects predicted by this scenario. Figure 20.4 below is one 'snapshot' map from this series, showing the depth and extent of inundation 17 hours after the initial breach, which was assumed to take place two hours before high tide. The figure shows the state of inundation after water from the second tidal cycle enters a terrain where many depressions are already quite full. Floodwater is predicted to quickly refill partially flooded storage cells and to spill over a railway embankment into a residential area. Staff in the operational headquarters at the Thames Barrier have evaluated these scenarios. The readily produced inundation maps were found extremely useful for approximating the amount of time available and for investigating possible strategies for evacuating residents in such circumstances. Given the short time available for making such decisions if a real event were to occur, the methods also have potential as a training tool for planning responses to other 'What If?' scenarios.

20.5 ADVANTAGES AND LIMITATIONS OF THE NEW METHODS

20.5.1 Advantages

In terms of data processing, the main advantage of the approach presented is that once the initial terrain pre-processing has been performed, which is a one-off requirement, the methods are able to produce results quickly and flexibly. These methods are not intended as a substitute for detailed hydraulic modelling. This is more of a rapid reconnaissance technique for assessing the approximate extent of flooding, which does not require the detailed calibration involved in setting up a hydraulic model for specific breach locations. The further potential for this approach is discussed briefly and implications considered under the following sub-headings:

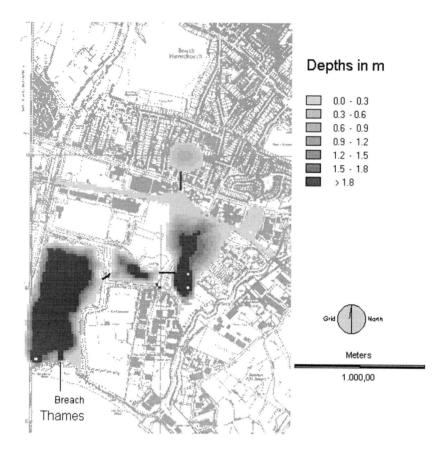

Figure 20.4 Status of inundation 17 hours after the breach (Background produced from 1993 Ordnance Survey map TQ58SW with the permission of the controller of Her Majesty's Stationery Office, © Crown Copyright ED/107A)

20.5.1.1   Real-time visualisation for the non-expert user

The simple model has the advantage of being extremely easy to use when the input files have been prepared. The user is able to interactively choose the location of the breach and to see the predicted extents and duration of flooding in faster than real time. Although the assumptions of the model have to be understood to assess the results critically, the capability to produce a range of flooding scenarios and to see immediately the geographical consequences can help to develop a better understanding of the likely effects of different types and locations of embankment breaches.

20.5.1.2 *Flood risk assessment can use more spatially distributed data*

To determine the extent of flooding, the main information needed in addition to water levels is the topography of the floodplain. For floods in natural rivers without embankments this task is often reduced to finding the area adjacent to a stretch of river with an elevation lower than the river stage of the event under consideration. This type of procedure is used to justify the costs of flood protection works in many countries. In Britain the Environment Agency (EA) estimates flood risk for the Ministry of Agriculture, Fisheries and Food (MAFF) as the area that would be flooded if there were no protection works (e.g. flood walls, embankments) on the river. This is not a very realistic assumption, as it over-estimates the extent of flooding and subsequent damage.

Even if a river overtops a long stretch of embankment, it is quite unlikely that the floodplain will be completely flooded. In the UK, river levels tend to decrease after only a few hours as the main flood passes or in case of tidal flooding because of the tidal cycle. The pattern of flooding resulting from a localised failure such as a riverbank breach will depend very much on the location of the breach. Such flooding is usually confined within embayments, into which the floodplain is structured by the river defence works, road embankments, walls and other man-made structures.

While a crude map of maximum flood extent, assuming no defensive works and an unlimited supply of floodwater can be prepared by hand on the basis of elevation contours, such a method is insensitive to the location of the breach and the structures on the floodplain. The methods developed here permit scenarios to be calculated for a number of breach locations, breach widths and water levels along a flood protection works. The readily produced inundation maps can then be combined with other data sets already available in digital form for infrastructure, land use, buildings and population figures, etc. to identify localities where the failure of a protection works is likely to produce the greatest damage.

20.5.1.3 *Potential to integrate algorithms within standard GIS systems*

The methods presented here for inundation modelling use data and pre-processing functions that are quite widely available within raster GIS. The algorithms use standard matrix operations and a straightforward pointer chain structure for keeping track of the connectivity of the storage depressions. These algorithms could be integrated within a standard GIS that already allows the generation of a local drain direction matrix and the delineation of watersheds. The model produces results in digital formats ready for immediate post-processing using standard GIS functionality, such as buffering and overlay analysis, allowing the query and reporting of flood predictions in full geographical context.

The methods that have been implemented are modular and relatively generic. For example, a different function for water input can be substituted for the breach flow model that is presently implemented. The FLOODFILL algorithm for spreading water over a terrain can be adapted to a range of related problems involving the distribution of water in a terrain that is structured into storage depressions.

20.5.2   Limitations

20.5.2.1   Handling multiple overflow points

The FLOODFILL algorithm for spreading water over a terrain is presently not able to handle overflow from one depression into more than one other. Within a depression, several points might exist with the same overflow height. The present method does not allow for more than one depression to be flooded at the same time. It is possible to allow for flow diversion but the question would immediately be how to distribute the flow. Only if the DTM resolution were high enough to derive realistic cross-sections around the overflow points could a reasonable method for partitioning the flow be implemented. In cases where the depression is bounded by an embankment of uniform height, the exact location of the overflow point is somewhat arbitrary. However, the overflow of water at one point along an embankment rather than at another, may affect which depressions beyond the bank subsequently flood.

20.5.2.2   Data availability

A general problem in the construction of digital terrain models for floodplains is that the required vertical resolution can usually not be obtained from existing map sources. To establish the curve of volumetric storage against water elevation for the interconnected depressions, to an accuracy comparable to a hydraulic model, a contour interval of 1 m or less is needed (Cunge *et al.*, 1980). Ordnance Survey 1:10,000 maps have a 5 m contour interval. 1:1250 mapping contains only individual spot heights. This problem will be overcome in the future as the need for detailed floodplain height information in urbanised areas and for low-lying coastal areas becomes widely accepted. In the UK, for example, the Environment Agency is undertaking a series of experiments using methods such as photogrammetry based on 1:5,000 scale aerial photography and use of LIDAR laser altimetry to create terrain models for major floodplains at sub-metre vertical resolution in the 0 – 10 m elevation range.

20.5.2.3   Regular grid data structure quite unsuited for flooding applications

The data structure of the DTM in this model is a regular two-dimensional grid. The cell size of this grid is the same within an area with large changes in elevation, as it is within an extensive flat area with little or no change in elevation. The features that control water flow in a floodplain are often of rather limited spatial extent. For example, a flood wall might have a width of 0.5 m. To ensure that the wall is represented correctly in the elevation model, the grid cell size has to be considerably smaller than 0.5 m. Otherwise, the resampling from the vector based triangular network into the grid will not always reproduce the crest of the wall as a continuous feature in the grid representation. The top of an embankment can appear to vary in height because some points will hit the crest while others will lie on the banks. As a correct representation of those linear features that enclose individual depressions are of major importance for the correct distribution of the inflowing water, errors of this kind should not appear in the terrain model. The present solution to maintaining the hydrological consistency of floodplain terrain models is to build these at very high spatial resolution, even though much of this detail is redundant, occupies large amounts of space and slows processing. Future research might investigate the potential for a variable resolution grid structure, where only important hydrological features are stored at high resolution.

20.6 CONCLUSIONS

A simple model has been developed for predicting the extents, depths and duration of floodwater inundation resulting from river embankment breaches. The approach taken in the modelling has been to adopt a relatively simple model of the process of floodwater inundation, but one that fully exploits high resolution digital terrain data to model the process by which surface depressions are likely to fill, coalesce and eventually drain. Initial results show that the model is able to provide plausible predictions in faster than real time and to produce maps of predicted flooding which could be used to respond to the immediate risk. Given the success of the initial pilot, we plan further work to verify the applicability of the simplifications in the current model and to test its robustness on terrain data derived from digital photogrammetry. This work will determine the feasibility of this approach for a real-time system for flood defence personnel to use in an emergency situation.

ACKNOWLEDGEMENTS

We are grateful to the Environment Agency for supporting this project, particularly to Andrew Bachelor and Christopher Richards from the Thames Barrier Operational HQ who assisted in formulating the breach scenario and in evaluating the scenarios predicted. Dick Greenaway and the staff of EA Survey Section kindly provided relevant mapping and height data for the test area. Thanks also to Willem van Deursen whose software PC RASTER was freely available to the project and who provided helpful comments on a draft of this chapter.

REFERENCES

Burrough, P.A., 1997, Environmental modelling with geographic information systems. In *Innovations in GIS 4*, Kemp, Z. (ed) (London: Taylor & Francis), pp. 143-153.

Cunge, J. A., Holly, F. M. and Verwey, A., 1980, *Practical Aspects of Computational River Hydraulics* (London: Pitman Publ. Ltd).

Deursen, W.P.A. van and Wesseling, C. G., 1993, *PC-RASTER, Manual Version 2,* (Utrecht: Utrecht University).

Evans, E. P. and Lany, P. H. von, 1983, A Mathematical Model of Overbank Spilling and Urban Flooding. In: *International. Conference on Hydraulic Aspects of Floods & Flood Control, 13-15 September, London, England.*

Fedra, K., 1996, Distributed models and embedded GIS: Integration strategies and case studies. In *GIS and Environmental Modelling: Progress and Research Issues,* Goodchild, M.F., Steyaert, L. T., Parks, B. O. Johnston, C, Maidment, D. R., Crane, M. and Glendinning, S. (eds) (Fort Collins: GIS World Inc.), pp. 413-417.

Gatrell, A. C. and Vincent, P., 1990, Managing Natural and Technological Hazards - The Role of GIS. *Regional Research Laboratory Initiative, Discussion Paper Number 7,* August 1990, (Sheffield: Sheffield University Press).

Goodchild, M.F., Steyaert, L. T., Parks, B. O. Johnston, C., Maidment, D. R., Crane, M. and Glendinning, S. (Eds), 1996, *GIS and Environmental Modelling: Progress and Research Issues*, (Fort Collins: GIS World Inc.).

Grayson, R.B., Bloschl, G., Barling, R. D. and Moore, I. D., 1993, Process, scale and constraints to hydrological modelling in GIS. In *Application of GIS to hydrology and water resources management,* Kovar, K. and Nachtnebel, H. P. (eds) IAHS Publication 211, pp.83-92.

Jensen, S. K. and Dominigue J. O., 1988, Extracting topographic structure from digital elevation data for GIS analysis. *Photogrammetric Engineering and Remote Sensing,* Vol. 54(11), pp. 1593-1600.

Johnston, C.A., 1998, *Geographic information systems in Ecology,* (London: Basil Blackwell).

Kovar, K. and Nachtnebel, H. P. (Eds), 1996, *Application of GIS in hydrology and water resources management* (HydroGIS '96), IAHS Publication 235.

Macmillan, R. A., Furley, P. A. and Healey, R.G., 1993, Using hydrological models and GIS to assist with the management of surface water in agricultural landscapes. In *Landscape Ecology and GIS*, Haines-Young, R., Green, D. R. and Cousins, S. H. (eds) (London: Taylor & Francis), pp. 181-210.

Moore, I. D., Turner, A.K., Wilson, J.P., Jenson, K. and Band, L.E., 1993, GIS and the land surface – subsurface process modelling. In *Environmental modelling with GIS*, Goodchild, M.F, Parks, B. O., and Steyaert, L. T. (eds) (New York: Oxford University Press), pp. 197-230.

Moore, I. D., 1996, Hydrological Modelling and GIS. In *GIS and Environmental Modelling: Progress and Research Issues*, Goodchild, M.F., Steyaert, L. T., Parks, B. O., Johnston, C., Maidment, D. R. Crane, M. and Glendinning, S. (eds) (Fort Collins: GIS World Inc.), pp. 143-148.

Raper, J. and Livingstone, D., 1994, Modelling environmental systems with GIS: theoretical barriers to progress. In *Innovations in GIS 1*, Worboys, M. (ed) (London: Taylor & Francis), pp. 229-240.

Roberts, T. and Anderson, J., 1994, Real Time Flood Forecasting with GIS. In *Current practices in modelling the management of stormwater impacts*, James, W. (ed) (Boca Raton, Florida, CRC Press,), pp. 281-291.

Wesseling, C. G., Karssenberg, D., Burrough, P.A. and van Deursen, W.P.A., 1996, Integrating dynamic environmental models and GIS: the development of a dynamic modelling language. *Transactions in GIS*, Vol. 1, pp. 40-48.

Wilson, J. P., 1996, GIS-based land surface/subsurface modelling: new potential for new models? In *Proceedings of the 3rd International Conference on Integrating GIS and Environmental Models*, Santa Fe, (Santa Barbara: National Centre for Geographical Information and Analysis).

Index